Gödel Without (Too Many) Tears

Gödel Without (Too Many) Tears

Peter Smith

LOGIC MATTERS

Published by Logic Matters, Cambridge

ISBN 978-1-916-90630-3 Hardback
ISBN 978-1-916-90631-0 Paperback

Additional resources for this publication at www.logicmatters.net.

Contents

Preface		vii
1	Incompleteness, the very idea	1
2	The First Theorem, two versions	11
3	Outlining a Gödelian proof	15
4	Undecidability and incompleteness	21
5	Two weak arithmetics	28
6	First-order Peano Arithmetic	40
7	Quantifier complexity	47
Interlude		51
8	Primitive recursive functions	54
9	Expressing and capturing the primitive recursive functions	66
10	The arithmetization of syntax	75
11	The First Incompleteness Theorem, semantic version	81
12	The First Incompleteness Theorem, syntactic version	87
Interlude		94
13	The Diagonalization Lemma	98
14	Undecidability	105
15	Rosser's Theorem	107
16	Tarski's Theorem	110
17	The Second Theorem and Hilbert's Programme	116
18	Proving the Second Incompleteness Theorem	123
19	A subtle difference	129
Further reading		133
Index of definitions		135

Preface

Why these lecture notes? After all, I have already written a rather long book, *An Introduction to Gödel's Theorems* (originally CUP, now freely downloadable). Surely that's more than enough to be going on with?

Ah, but there's the snag. It *is* more than enough. In the writing, as is the way with these things, that book grew far beyond the scope of the original notes on which it was based. And while I hope the result is still quite accessible to someone prepared to put in the required time and effort, there is – to be frank – a *lot* more in the book than is really needed by those wanting a first encounter with the famous incompleteness theorems.

Quite a few readers might therefore appreciate a cut-down version of some of the material in the book – an introduction to the *Introduction*, if you like. Hence *Gödel Without (Too Many) Tears*. There are occasional footnotes referring to sections of the book, indicating where topics are discussed further: but you don't have to chase up those references to get a more limited but still coherent story in these notes.

There isn't much purely philosophical discussion here: the aim, rather, is to put you in a position where you have a secure enough initial understanding of what's going on logically that you can then sensibly make a start on thinking about any (supposed) philosophical implications.

So what background in logic do we presuppose? What do you need to bring to the party? Very little. If you have done a standard introductory logic course, and have the patience to follow some simple mathematical arguments, you should have little difficulty in following the exposition here. I have given proofs of most of the important theorems stated in these notes, especially when the proofs involve some neat ideas. But I have left a few proofs for enthusiasts to follow up elsewhere, if trekking through the details has little intrinsic interest.

A first version of these notes – call it *GWT1* – was written to accompany the last outings of a short lecture course given in Cambridge (which was also repeated at the University of Canterbury, NZ). Many thanks to many students for useful feedback.

GWT1 aimed to bridge the gap between classroom talk'n'chalk which just highlighted the Really Big Ideas, and the much more detailed treatments of topics available in my book. However, despite that intended role, I did try to make *GWT1* reasonably stand-alone.

Those notes were tied to the first edition of my book, *IGT1*, as published in 2007. A significantly improved second edition of the book, *IGT2*, was published in 2013. So I updated *GWT1* in 2014 to make a corresponding second version of the notes – call it *GWT2*.

It's time to revisit the notes, and make some more improvements. Rewriting has been occupational therapy for me in the time of pandemic, and I hope the result will be equally distracting for those who like learning about logic. So here is *GWT3*.

I should particularly thank Henning Makholm for comments on an earlier version of *GWT*, and David Auerbach, Sam Butchart, David Furcy, David Makinson and Rowsety Moid for more recent comments. But many others too have at various stages kindly let me know about typos and more serious mistakes, or made helpful suggestions. I am very grateful to everyone!

1 Incompleteness, the very idea

1.1 A brief note on Kurt Gödel

By common agreement, Kurt Gödel (1906–1978) was the greatest logician of the twentieth century. Born in what is now Brno, and educated in Vienna, Gödel left Austria for the USA in 1940, and spent the rest of his life at the Institute for Advanced Study at Princeton.

Gödel's doctoral dissertation, written when he was 23, established the *completeness* theorem for the first-order predicate calculus (showing that a standard proof system for first-order logic indeed captures all the semantically valid inferences).

Later he would do immensely important work on set theory, as well as make seminal contributions to proof theory and to the philosophy of mathematics. He even wrote on models of General Relativity with 'closed timelike curves' (where, in some sense, time travel is possible). Always a perfectionist, he became a very reluctant publisher (some of his philosophically most interesting work is in the volume of Unpublished Essays and Lectures in his *Collected Works*).

Talk of 'Gödel's Theorems' typically refers, however, to the two *incompleteness* theorems presented in an epoch-making 1931 paper. And it is these theorems, and more particularly the First Theorem, that these notes are all about. (Yes, that's right: Gödel proved a 'completeness theorem' and also 'incompleteness theorems'. We'll explain the difference in a moment.)

The impact of the incompleteness theorems on foundational studies is hard to exaggerate. For, putting it crudely and a little bit tendentiously, they sabotage the ambitions of two major foundational programs – logicism and Hilbert's Programme. We'll say just a little about logicism in this chapter, and something about Hilbert's Programme much later, when we get round to discussing the Second Theorem in Chapter 17. But you don't have to know anything about this background to find the two theorems intrinsically fascinating.

1.2 The idea of an axiomatized formal theory

The title of Gödel's great 1931 paper translates as *'On formally undecidable propositions of* Principia Mathematica *and related systems I'*.

The 'I' here indicates that it was intended to be the first part of what was

1

going to be a two part paper, with Part II spelling out the proof of the Second Theorem which is only very briefly indicated in Part I. But Part II was never written. We'll see in due course why Gödel thought he didn't need to bother.

This title itself gives us a number of things to explain. What's a 'formally undecidable proposition'? What is *Principia Mathematica*? Ok, you've probably heard of that triple-decker work by A. N. Whitehead and Bertrand Russell, now more than a century old and very little read except by historians of logic: but what is the project of that book? And what counts as a 'related system' – a 'system' suitably related, that is, to the one in *Principia*? In fact, just what is meant by 'system' here?

Let's take the last question first. We will take a 'system' (in the relevant sense) to be an *effectively axiomatized formal theory*.[1] But what does that mean?

The general idea of an axiomatized formal theory is no doubt familiar. Roughly: you fix on a formalized language, set down some axioms stated in that language, specify some apparatus for formally deriving results from your axioms, and there you have a theory. But now we need to be more explicit: our focus is going to be on theories which, in headline terms, have

   (i) an effectively formalized language,
  (ii) an effectively decidable set of axioms,
 (iii) an effectively formalized proof system.

We'll explain these headlines in just a moment. But first, the new idea you need to get your head round here is the intuitive notion of *effective decidability*.

Let's say, as a first shot:

Defn. 1. *A property P (defined over some domain of objects D) is* effectively decidable *iff[2] there's an algorithm (a finite set of instructions for a deterministic computation) for settling in a finite number of steps, for any object $o \in D$, whether o has property P.*

Likewise, a set Σ is effectively decidable *iff the property of being a member of that set is effectively decidable.*

Relatedly, the answer to a question Q is effectively determinable *iff there is an algorithm which gives the answer (again by a deterministic computation, in a finite number of steps).*

To put it another way, a property is effectively decidable just when there's a step-by-step mechanical routine for settling whether o has property P, such that a suitably programmed deterministic computer could in principle implement the routine (idealizing away from practical constraints of time, etc.). Similarly, the answer to a question is effectively determinable just when a suitably programmed computer could deliver the answer (in principle, in a finite time).

---

[1] It will turn out that Gödel originally had in mind just a central subclass of systems in this wide sense; but let's not complicate the story yet.

[2] 'Iff' is of course the logician's abbreviation for 'if and only if'.

Two initial examples from elementary logic: the property of being a tautology is effectively decidable (by a truth-table test); and we can effectively determine what is the main connective of a sentence (by bracket counting).

How satisfactory are our definitions, though? We have just invoked the idea of what an idealized computer could (in principle) do by implementing some algorithm. That idea surely needs further elaboration. It turns out, however, that the notion of effective decidability is actually very robust: what is algorithmically-computable-in-principle according to one sensible sharpened-up definition is exactly what is algorithmically-computable-in-principle according to any other sensible sharpened-up definition. Of course, it's not at all trivial that this is how things are going to pan out. So for the moment you are going to have to take it on trust (sorry!) that Defn. 1 *can* be put into good shape by sharpening the notion of effective decidability.

Against this background, we can now explain those conditions (i) to (iii) for being an effectively axiomatized formal theory.

(i) We'll assume that the basic idea of a *formalized language L* is familiar from earlier logic courses. But note, a language, for us, has both a *syntax* and an intended *semantics*:

1. The syntactic rules fix which strings of symbols form terms, which form wffs (i.e. well-formed formulas), and in particular which strings of symbols form sentences, i.e. closed wffs with no unbound variables dangling free.

2. The semantic rules assign unique interpretations, i.e. assignments of truth-conditions, to every sentence of the language.

It is not at all unusual for logicians to call a system of uninterpreted strings of symbols a 'language'. But I really think we should deprecate that usage. Sometimes below I'll talk about an 'interpreted' language for emphasis: but strictly speaking – in my idiolect – that's redundant.

The familiar way of presenting the syntax of a formal language is by (a) first specifying some finite set of symbols,[3] and then giving rules for building up more and more complex expressions from these symbols. And we standardly do this in such a way that (b) we can *effectively* decide whether a given string of symbols counts as a term, or a wff, or a wff with one free variable, or a sentence (we can give algorithms which decide well-formedness, etc.).

The familiar way of presenting the semantics is then to assign semantic values to the basic non-logical expressions of the language, fix domains of quantification, and then give rules for working out the truth-conditions of longer and longer expressions in terms of the way they are syntactically built up from their parts. In a standard formal language, we can *effectively* recover from a sentence its 'constructional history', i.e. mechanically determine the way it is syntactically

---

[3] "Finite? But might we not need an unlimited, potentially infinite, supply of variables, for example?" Sure. But we can build up an infinite list of variables from finite resources, as in 'x, x′, x″, x‴, . . .'. We lose no relevant generality for our purposes in keeping our basic symbol-set finite.

built up from its parts; then, relying on this information, (c) we can use the semantic rules to mechanically work out the interpretation of any given sentence. (Read that carefully! What we can mechanically work out is what the sentence *says*. But it is one thing to work out the *conditions* under which a sentence is true, and – usually – something quite different to work out whether those conditions are met, i.e. work out whether the sentence actually is *true*.)

Let's say that a formalized language which shares these characteristics (a), (b) and (c) is effectively formalized. So, in sum,

Defn. 2. *An interpreted language L is* effectively formalized *iff (a) it has a finite set of basic symbols, (b) syntactic properties such as being a term of the language, being a wff, being a wff with one free variable, and being a sentence, are effectively decidable and the syntactic structure of any sentence is effectively determinable, and (c) this syntactic structure together with the semantic rules can be used to effectively determine the unique intended interpretation of any sentence.*

Why do we want (b) the syntactic properties of being a sentence, etc., to be effectively decidable? Well, the point of setting up a formal language is, for a start, to put issues of what is and isn't e.g. a sentence beyond dispute, and the best way of doing that is to ensure that even a suitably programmed computer could decide whether a string of symbols is or is not a sentence of the language. Why do we want (c) the unique truth-conditions of a sentence to be effectively determinable? Because we don't want any ambiguities or disputes about interpretation either.

(ii) A theory is sometimes defined to be just any old set of sentences. We are concerned, though, with the narrower notion of an *axiomatized theory*. We highlight some bunch of sentences Σ as giving *axioms* for the theory T; we give T some *proof system*, i.e. some deductive apparatus; and then all the sentences that are derivable from axioms in Σ using the deductive apparatus are T's *theorems*.

But what does it take for T to be an *effectively* axiomatized theory, apart from its using an effectively formalized language? For a start, we require it to have an effectively decidable set of axioms. Why? Because if we are in the business of pinning down a theory by axiomatizing it, then we will normally want to avoid any possible dispute about what counts as a legitimate starting point for a proof by ensuring that we can mechanically decide whether a given sentence is indeed one of the axioms.

(iii) But just laying down a bunch of axioms would be pretty idle if we can't deduce conclusions from them! An axiomatized theory T will, as we said, come equipped with a deductive proof system, a set of rules for deriving further theorems from our initial axioms. But a proof system such that we couldn't routinely tell whether its rules are being followed again wouldn't have much point for practical purposes. Hence we naturally also require that a theory has an effectively formalized proof system, i.e. one where it is effectively decidable whether a given array of wffs is a well-constructed derivation from the axioms according to the rules of the proof system.

Note, it doesn't matter for our purposes whether the proof system is an axiomatic logic, a natural deduction system, a tree/tableau system, or a sequent calculus – so long as it is effectively checkable that a candidate proof-array has the property of being properly constructed according to the rules of the proof system.

So, in summary of (i) to (iii),

Defn. 3. *An* effectively axiomatized *formal theory T has an effectively formalized language L, a certain class of L-wffs are picked out as axioms where it is effectively decidable what's an axiom, and it has a proof system such that it is effectively decidable whether a given array of wffs is a derivation from the axioms according to the rules.*

Careful, though! To say that it must be effectively decidable whether a candidate T-proof of φ is indeed a kosher proof is not, repeat *not*, to say that it must be effectively decidable whether φ actually *has* a T-proof.

To stress the point: it is one thing to be able to effectively *check* that some proposed proof follows the rules; it is another thing to be able to effectively *decide in advance* whether there exists a proof waiting to be discovered. (Looking ahead, we will see as early as Chapter 4 that any formal effectively axiomatized theory T containing a modicum of arithmetic is such that, although you can mechanically check a purported proof of φ to see whether it *is* a proof, there's no mechanical way of telling of an arbitrary φ whether it is provable in T or not.)

Henceforth, when we talk about formal theories, we will be concerned with effectively axiomatized theories (unless we explicitly say otherwise).

1.3 'Formally undecidable propositions' and negation incompleteness

Some familiar logical notation, applied to formal theories:

Defn. 4. *'T $\vdash \varphi$' says: there is a formal derivation in T's proof system from T-axioms to the sentence φ as conclusion (in short, φ is a T-theorem).*

Defn. 5. *'T $\vDash \varphi$' says: any way of (re)interpreting the non-logical vocabulary that makes all the axioms of T true makes φ true.*

So '$\vdash$' officially signifies *derivability* in T, which is a syntactically-definable relation. While '$\vDash$' signifies *logical entailment*, a semantic relation defined by generalizing over interpretations.[4]

---

[4] "You said that a theory's language has a built-in interpretation: now you seem to be forgetting that." Not at all. Recall, some premises *logically* entail a given conclusion iff the inference from the premises to the conclusion is necessarily truth-preserving *just in virtue of the distribution of logical operators in the relevant sentences.* That's why our official

Of course, we normally want a formal derivation to be truth-preserving; so we will want our proof system to respect logical entailments, requiring that if $T \vdash \varphi$ then indeed $T \vDash \varphi$. In a word, we require the proof system used in a sensible theory to be *sound*.

We can't in general insist on the converse. Not every relation of logical entailment can be captured in a proof system: so there can a theory T where we don't have $T \vdash \varphi$ whenever $T \vDash \varphi$. But take the very important special case where the theory T has a standard first-order logic. In a classical first-order setting, if an inference from T to φ is semantically valid, i.e. is necessarily truth-preserving, then there will indeed be a formal deduction of φ from the axioms of T. This was shown for a Hilbert-style axiomatic deductive system by Gödel in his doctoral thesis: hence *Gödel's completeness theorem*.

Some more key definitions. We will be interested in what claims a theory T can settle, one way or the other. So we say

Defn. 6. *If T is a theory, and φ is some sentence of the language of that theory, then T formally decides φ iff either $T \vdash \varphi$ or $T \vdash \neg\varphi$.*

Hence,

Defn. 7. *A sentence φ is* formally undecidable *by T iff $T \nvdash \varphi$ and $T \nvdash \neg\varphi$.*

A related bit of terminology:

Defn. 8. *A theory T is* negation complete *iff it formally decides every closed wff of its language – i.e. for every sentence φ, $T \vdash \varphi$ or $T \vdash \neg\varphi$.*

So there are formally undecidable propositions in a theory T if and only if T isn't negation complete.

It might help to fix ideas, and distinguish the two notions of completeness, if we take a toy example. Suppose theory T is built in a propositional language with just three propositional atoms, p, q, r, plus the usual propositional connectives. We give T a standard propositional classical logic (pick your favourite flavour of system!). And assign T just a single non-logical axiom: $(p \wedge \neg r)$.

Then, by assumption, T has a *semantically-complete logic*, since standard propositional calculi are complete. That is to say, for any wff φ of T's limited language, if $T \vDash \varphi$, i.e. if T tautologically entails φ, then $T \vdash \varphi$.

However, trivially, T is not a *negation-complete theory*. For example T can't decide whether q is true. And there are lots of other wffs φ for which both $T \nvdash \varphi$ and $T \nvdash \neg\varphi$.

Our toy example shows that it is very, very easy to construct negation-incomplete theories with formally undecidable propositions: just hobble your theory T by leaving out key assumptions about the matter in hand!

definition of logical entailment abstracts away from the given meaning of the non-logical constituents of the sentences while keeping the meaning of the logical apparatus fixed, and why we generalize over all possible reinterpretations of the non-logical constituents.

But suppose we are trying to fully pin down some body of truths (e.g. the truths of basic arithmetic) using a formal theory T. We fix on an interpreted formal language L apt for expressing such truths. And then we'd ideally like to lay down enough axioms framed in L such that, for any L-sentence φ, then $T \vdash \varphi$ just when φ is true. So, making the classical assumption that for any sentence φ, either φ is true or $\neg\varphi$ is true, we'd very much like T to be such that either $T \vdash \varphi$ or $T \vdash \neg\varphi$ (but not both).

In other words, it is natural to aim for theories T which are indeed negation complete.

1.4   Seeking a negation-complete theory of arithmetic

The elementary arithmetic of addition and multiplication is child's play (literally!). So we should be able to wrap it up in a nice formal theory, aiming for negation completeness.

Let's first fix on a formal *language of basic arithmetic* designed to express elementary arithmetical propositions. We will give this language

(i) a term '0' to denote zero; and
(ii) a sign 'S' for the successor function (the 'next number') function.

This means that we can construct the sequence of terms '0', 'S0', 'SS0', 'SSS0', ... to denote the natural numbers 0, 1, 2, 3, ... These are our language's *standard numerals*, and by using a standard numeral our language can denote any particular number.

We will also give this language

(iii) function signs for addition and multiplication, plus
(iv) the usual first-order logical apparatus, including the identity sign: quantifiers are interpreted as running over the natural numbers.

(We aren't building in subtraction and division as primitives, however. But subtraction is definable in terms of addition, formalizing the idea that $n - m$ is the number k such that $m + k = n$, if there is such a number. And similarly division is definable in terms of multiplication.)

Now, it is entirely plausible to suppose that, whether or not the answers are readily available to us, questions posed in this language of basic arithmetic have entirely determinate answers. Why? Well, take the following two bits of data:

(a) The fundamental zero-and-its-successors structure of the natural number series.
(b) The nature of addition and multiplication as given by the school-room explanations.

By (a) we mean that zero is not a successor, every number has a successor, distinct numbers have distinct successors, and so the sequence of zero and its successors never circles round but marches on for ever: moreover there are no strays – i.e. every natural number is in that sequence starting from zero. We will say more about (b) in due course. But it is very plausible to suppose that (a) and (b) together should fix the truth-value of every sentence of the language of basic arithmetic (after all, what more could it take?).

But (a) and (b) seem so very basic and straightforward. So we will surely expect to be able to set down some axioms which characterize (a) the number series, and (b) addition and multiplication: in other words, we should surely be able to frame axioms which codify what we teach the kids. And then the thought that (a) and (b) fix the truths of basic arithmetic becomes the thought that our axioms capturing (a) and (b) should settle every such truth. In other words, if φ is a true sentence of the language of successor, addition, and multiplication, then φ is provable from our axioms (and if φ is a false sentence, then $\neg\varphi$ is provable).

In sum, whatever might be the case with fancier realms of mathematics, it is very natural to suppose that we should at least be able to set down a negation complete (and effectively axiomatized) theory of basic arithmetic.

1.5 Logicism and *Principia*

It is natural to ask: what could be the *status* of the axioms of a formal theory of arithmetic – e.g. the status of the formalized version of a truth like 'every number has a unique successor'? That hardly looks like a mere empirical generalization (something that could in principle be empirically refuted).

I suppose you might be a Kantian who holds that the axioms encapsulate 'intuitions' in which we grasp the fundamental structure of the numbers and the nature of addition and multiplication, where these 'intuitions' are a special cognitive achievement in which we somehow represent to ourselves the arithmetical world.

But talk of such intuitions is, to say the least, puzzling and problematic. So we could very well be tempted instead by Gottlob Frege's seemingly more straightforward view that the axioms are *analytic*, simply truths of logic-plus-definitions. On this view, we don't need Kantian 'intuitions' going beyond logic: logical reasoning from definitions is enough to get us the axioms of arithmetic, and more logic gives us the rest of the arithmetic truths from these axioms. This Fregean line is standardly dubbed *logicism.*

If this is to be more than wishful thinking, we need a well-worked-out logical system within which to pursue a logicist derivation of arithmetic. Famously, and to his eternal credit, Frege gave us the first competent system of quantificational logic in his *Begriffsschrift* of 1879. But equally famously, Frege's own attempt to be a logicist about basic arithmetic (in fact, for him, about more than basic arithmetic) hit the rocks, because – as Russell showed – the full deductive proof system that he later used, going beyond core quantificational logic, is inconsistent in a pretty elementary way. Frege's full system is beset by Russell's Paradox.[5]

---

[5]Roughly, Frege's full system implies that there is a set of all sets which are not members of themselves – does that set belong to itself?

That devastated Frege, but Russell himself was undaunted. Still gripped by logicist ambitions he wrote:

> All mathematics [yes! – *all* mathematics] deals exclusively with concepts definable in terms of a very small number of logical concepts, and ... all its propositions are deducible from a very small number of fundamental logical principles.

That's a huge promissory note in Russell's *The Principles of Mathematics* (1903). And *Principia Mathematica* (three volumes, though unfinished, 1910, 1912, 1913) is Russell's attempt with Whitehead to start making good on that promise.

The project of *Principia*, then, is to set down some logical axioms and definitions in which we can deduce, for a start, all the truths of basic arithmetic (so giving us a negation-complete theory at least of arithmetic). Famously, the authors eventually get to prove that $1 + 1 = 2$ at *110.643 (Volume II, page 86), accompanied by the wry comment, 'The above proposition is occasionally useful'. So far so good! But can Russell and Whitehead, in principle, prove *every* truth of arithmetic?

1.6   Gödel's bombshell

Principia, frankly, is a bit of a mess – in terms of clarity and rigour, it's quite a step backwards from Frege. And there are technical complications which mean that not all *Principia*'s axioms are clearly 'logical' even in a stretched sense. In particular, there's an appeal to a brute-force *Axiom of Infinity* which in effect stipulates that there is an infinite number of objects; and then there is the notoriously dodgy so-called *Axiom of Reducibility*. But we don't need to go into details; for we can leave those worries aside – they pale into insignificance compared with the bombshell exploded by Gödel.

For Gödel's First Incompleteness Theorem sabotages not just the grand project of *Principia*, but shows that *any* attempt to pin down *all* the truths of basic arithmetic in a theory with nice properties like being effectively axiomatized is in fatal trouble. His First Theorem says – at a rough first shot – that *nice theories containing enough arithmetic are always negation incomplete*: for any nice theory T, there will be arithmetic truths that can't be proved in that particular theory.

A moment ago, it didn't seem at all ambitious to try to capture all the truths of basic arithmetic in a single (consistent, effectively axiomatized) theory. But attempts to do so – and in particular, attempts to do this in a way that would appeal to Frege and Russell's logicist instincts – must always fail. Which is a rather stunning result![6]

---

[6]'Hold on! I've heard of neo-logicism which has its enthusiastic advocates. How can that be so if Gödel showed that logicism is a dead duck?' Well, we might still like the idea that some logical principles plus what are more-or-less definitions (in a language richer than that of first-order logic) together *semantically* entail all arithmetical truths, while allowing that we

How did Gödel prove his result? Well, let's pause for breath; the next chapter explains more carefully what the theorem (in two versions) claims, and then in Chapter 3 we outline a Gödelian proof of one version.

---

can't capture the relevant semantic entailment relation in a single effectively axiomatized deductive system of logic. Then the resulting overall system of arithmetic won't count as a formal effectively axiomatizable theory, and Gödel's theorems don't straightforwardly apply. But all that is another story.

2 The First Theorem, two versions

2.1 Soundness, consistency, etc.

Let's read into the record two standard definitions:

Defn. 9. *A theory T is* sound *iff its axioms are true (on the interpretation built into T's language), and its proof system is truth-preserving, so all its theorems are true.*

Defn. 10. *A theory T is (syntactically)* consistent *iff there is no φ such that $T \vdash \varphi$ and $T \vdash \neg\varphi$, where '$\neg$' is T's negation operator.*

In a classical setting, if T is inconsistent, then $T \vdash \varphi$ for all φ. So another way of defining consistency is by saying that T is consistent iff for some φ, $T \nvdash \varphi$. And of course, soundness implies consistency. We shouldn't need to delay over these no doubt familiar ideas.

But we also need another (quite natural) definition to use in this chapter:

Defn. 11. *The formalized interpreted language L* contains the language of basic arithmetic *if L has a term which denotes zero and function symbols for the successor, addition and multiplication functions defined over numbers – these can be either built-in as primitives or introduced by definition – and has the usual connectives, the identity predicate, and can express quantifiers running over the natural numbers.*

An example might be the language of set theory, in which we can define zero, successor, addition and multiplication in standard ways, and express restricted quantifiers running over just zero and its successors.

(OK, you might worry whether the natural number system referred to in set theory is the genuine article or just a structurally equivalent surrogate. But then what is 'the genuine article'? We are not going to tangle with *that* messy issue, as we have quite enough other things to worry about! When we talk of quantifying over numbers inside e.g. set theory, then, understand that to be quantifying either over natural numbers or over whatever surrogates we can take to play the role of natural numbers. Nothing relevant to our project hangs on the difference.)

2.2   Two theorems distinguished

In his 1931 paper, Gödel proves (more or less) the following:[1]

Theorem 1. *Suppose T is a formal axiomatized theory whose language contains the language of basic arithmetic. Then, if T is sound, there will be a true sentence G_T of basic arithmetic such that $T \nvdash G_T$ and $T \nvdash \neg G_T$, so T is negation incomplete.*

We will outline a pivotal part of Gödel's proof in the next chapter.

However this version of an incompleteness theorem *isn't* what is most commonly referred to as *the* First Theorem, nor is it the result that Gödel foregrounds in his 1931 paper. For note, Theorem 1 tells us what follows from a *semantic* assumption, namely the assumption that T is sound. And soundness is defined in terms of truth.

Now, post-Tarski, most of us aren't particularly scared of the notion of the truth. To be sure, there are issues about how best to treat the notion formally, to preserve as many as possible of our pre-formal intuitions while e.g. blocking the Liar Paradox. But most of us don't regard the relevant notion of a sound theory as metaphysically loaded in an obscure and worrying way. However, Gödel was writing at a time when, for various reasons (think logical positivism!), the very idea of truth-in-mathematics was under some suspicion. It was therefore *extremely* important to Gödel that he could show that you don't need to deploy any semantic notions to get an incompleteness result. So he goes on to demonstrate (more or less) a result which we can put schematically like this:

Theorem 2. *Suppose T is a formal axiomatized theory whose language contains the language of basic arithmetic. Then, if T is consistent and can prove a certain modest amount of arithmetic (and has a certain additional property that any sensible formalized arithmetic will share), there will be a sentence G_T of basic arithmetic such that $T \nvdash G_T$ and $T \nvdash \neg G_T$, so T is negation incomplete.*

Being consistent is a syntactic property; being able to formally prove enough arithmetic is another syntactic property; and the mysterious additional property which I haven't explained is syntactically defined too. So *this* version of the incompleteness theorem only makes syntactic assumptions.

Of course, we'll need to be a lot more explicit about the details in due course; but this indicates the general *character* of Gödel's central result. Our 'can prove a certain modest amount of arithmetic' gestures at what it takes for a theory to be sufficiently related to *Principia*'s for the theorem to apply (recall the title of the 1931 paper). But I'll not pause here to spell out just how much arithmetic that is, though we'll eventually find that it is stunningly little. Nor will I pause to explain that 'additional property' condition. We'll meet it in due course, but

---

[1] I say 'more or less' because, as footnoted in §1.2, Gödel's idea of a formalized theory was in fact narrower than our notion of an effectively axiomatized theory: we'll explain the difference, and explain why it doesn't really matter, later.

also eventually see how – by a cunning trick discovered by J. Barkley Rosser in 1936 – we can drop that extra condition.

For now, then, the first take-away message of this chapter is that the incompleteness theorem does come in two flavours. There's a version making a *semantic* assumption (the relevant theory T needs to expressively rich enough and sound), and there's a version making only a *syntactic* assumption (about what T can derive from its axioms). It is important to keep this firmly in mind.

2.3   Incompleteness and incompletability

Let's concentrate for the moment on the first, semantic, version of the First Theorem.

Suppose, then, that T is a sound theory which contains the language of basic arithmetic. Then, the claim is, we can find a true G_T such that $T \nvdash \mathsf{G}_T$ and $T \nvdash \neg\mathsf{G}_T$. Let's be really clear: this doesn't, repeat *doesn't*, say that G_T is 'absolutely unprovable', whatever that could mean. It just says that G_T and its negation are unprovable-in-T.

Ok, you might well ask, why don't we simply 'repair the gap' in T by adding the true sentence G_T as a new axiom? Well, consider the theory $U = T + \mathsf{G}_T$ (to use an obvious notation). Then (i) U is still sound, since the old T-axioms are true, the added new axiom is true, and the theory's logic is still truth-preserving. (ii) U is still a properly formalized theory, since adding a single specified axiom to T doesn't make it undecidable what is an axiom of the augmented theory. (iii) U's language still contains the language of basic arithmetic. So Theorem 1 still applies, and we can find a sentence G_U such that $U \nvdash \mathsf{G}_U$ and $U \nvdash \neg\mathsf{G}_U$. And since U is stronger than T we have, a fortiori, $T \nvdash \mathsf{G}_U$ and $T \nvdash \neg\mathsf{G}_U$. In other words, 'repairing the gap' in T by adding G_T as a new axiom leaves some other sentences that were undecidable in T *still* undecidable in the augmented theory.

And so it goes. Keep throwing more and more additional true axioms at T and our theory will remain negation-incomplete, unless it stops being effectively axiomatized. So here's the second important take-away message of the chapter: when the conditions for Theorem 1 apply, then the theory T will not just be incomplete but in a good sense T will be *incompletable*.[2] (We'll see in due course that just the same holds when the conditions for Theorem 2 apply.)

So we should perhaps really talk of the First *Incompletability* Theorem.

2.4   The completeness and incompleteness theorems

We have already emphasized the distinction we need in §1.3, and we illustrated it then with a toy example. But experience suggests that it will do no harm at

---

[2]Suppose we take a language of basic arithmetic, and take *all* the true sentences of the language as axioms. Then yes, by brute force, we get a negation-complete theory! What Theorem 1 will tell us is that this theory can't be an effectively axiomatized theory – meaning that we can't effectively decide what's a true sentence of the language.

all to repeat the point! So: a semantic completeness theorem of the kind you are no doubt familiar with from elementary logic is about the relation between semantic and syntactic consequence relations. For example:

> If T is a theory cast in a first-order language with a standard first-order deductive apparatus, then for any φ, if $T \vDash \varphi$ then $T \vdash \varphi$.

(That's Gödel's completeness theorem.) But it doesn't follow from T's having a *complete logic* in this sense that T is a negation-complete *theory*. For example, if T is a theory of arithmetic with a standard first-order deductive apparatus it certainly has a semantically complete logic; but it can easily be a negation-incomplete theory. Just miss out axioms for addition (say), and there can be lots of wffs φ (those involving addition) such that neither $T \vdash \varphi$ nor $T \vdash \neg\varphi$!

Of course, that's a *very* boring way of being negation incomplete. And, as we said before, we might reasonably have expected that such incompleteness can always be repaired by judiciously adding in the needed axioms. The First Incompleteness Theorem tells us, however, that try as we might, every theory of arithmetic satisfying certain desirable conditions (even if it has a semantically complete logic) must *remain* negation incomplete as a theory.

3 Outlining a Gödelian proof

3.1 A notational convention

Before continuing, we should highlight a notational convention that we have already been using and which we will continue to use throughout these notes:

1. Expressions in informal mathematics will be in ordinary serif font, with variables, function letters etc. in italics. Examples:

$$2 + 1 = 3,\; n + m = m + n,\; S(x + y) = x + Sy.$$

2. Particular expressions from formal systems – and abbreviations of them – will be in sans serif type. Examples:

$$\mathsf{SSS0},\; \mathsf{SS0 + S0 = SSS0},\; \exists x\, x = 0,\; \forall x \forall y (x + y = y + x).$$

3. Greek letters, like 'Σ' and 'φ', are schematic variables in the metalanguage (so, in our case, they are added to logicians' English), which we can use e.g. in generalizing about wffs of our formal systems.

There will be a lot of to-and-fro in our discussions between claims of informal mathematics, samples of formal expressions and formal proofs, and general claims about formal proofs. It is essential to be clear which is which, and our notational convention should help considerably.

3.2 Formally expressing numerical properties, relations and functions

In the next few sections, we are going to prepare the ground for an outline sketch of how Gödel proved (a version of) Theorem 1.

We start with a couple more definitions. Recall, we said that the standard numerals of a language of basic arithmetic are the expressions '0', 'S0', 'SS0', 'SSS0', ... Let's now introduce a handy notational convention:

Defn. 12. *We will use '$\overline{n}$' to abbreviate the numeral denoting the number* n.

So '$\overline{n}$' will consist of n occurrences of 'S' followed by '0'. Hence '$\overline{5}$' abbreviates 'SSSSS0' which in a formal language with standard numerals denotes what '5' denotes in ordinary language.

Now assume we are dealing with a language L which has standard numerals (we'll also assume L has the usual apparatus of variables). Then we will say:

Defn. 13. *The open wff $\varphi(\mathsf{x})$ of the language L expresses the numerical property P just when $\varphi(\overline{\mathsf{n}})$ is true iff n has property P. Similarly, the formal wff $\psi(\mathsf{x}, \mathsf{y})$ expresses the numerical two-place relation R just when $\psi(\overline{\mathsf{m}}, \overline{\mathsf{n}})$ is true iff m has relation R to n. And the formal wff $\chi(\mathsf{x}, \mathsf{y})$ expresses the numerical one-argument function f just when $\chi(\overline{\mathsf{m}}, \overline{\mathsf{n}})$ is true iff $f(m) = n$.*

Hopefully, this definition should seem entirely natural.[1] For example, the wff $\exists \mathsf{y}\, \mathsf{x} = (\mathsf{y} + \mathsf{y})$ expresses the property of being an even number. Why? Because $\exists \mathsf{y}\, \overline{\mathsf{n}} = (\mathsf{y} + \mathsf{y})$ is true just in case n is the sum of some natural number with itself, i.e. is twice some number.

Note, as we have defined it, for a wff to express the property of being an even number is just for it to be true of the even numbers, i.e. just for the wff to have the right *extension*. For example, take the open wffs $\exists \mathsf{y}\, \mathsf{x} = ((\mathsf{S0} + \mathsf{S0}) \times \mathsf{y})$ and $\exists \mathsf{y}(\mathsf{x} = (\mathsf{y} + \mathsf{y}) \wedge \mathsf{S0} + \mathsf{S0} = \mathsf{SS0})$. These differ in intuitive sense, but again are satisfied by just the even numbers, so also count as expressing the property of being even. The same point holds more generally: expressing is just a matter of having the right extension.

Though we won't need it, the generalization of our definition to cover expressing many-place relations and many-argument functions is obvious enough.

3.3 Gödel numbers

And now for a key new idea. These days, we are entirely familiar with the fact that all kinds of data can be coded up using numbers: the idea was perhaps not in such everyday currency in 1931. But even then, the following sort of definition should have looked quite unproblematic:

Defn. 14. *A Gödel-numbering scheme for a formal theory T is some effective way of coding expressions of T (and sequences of expressions of T) as natural numbers. The scheme gives an algorithm for sending an expression (or sequence of expressions) to a number; and an algorithm for undoing the coding, sending a code number back to the expression (sequence of expressions) it codes. Relative to a choice of scheme, the code number for an expression (or a sequence of expressions) is its unique Gödel number.*

---

[1] Fine print. '$\varphi(\mathsf{x})$' indicates, of course, a wff with one or more occurrences of the variable 'x' free. The particular choice of free variable doesn't matter: '$\varphi(\mathsf{y})$', for instance, will do as well as '$\varphi(\mathsf{x})$'. '$\varphi(\overline{\mathsf{n}})$' is the result of replacing occurrences of the variable 'x' in '$\varphi(\mathsf{x})$' by the standard numeral for n. As you knew!

If you've been well brought up, you will prefer to use the symbolism '$\varphi(\xi)$', using a place-holding metavariable to mark a gap, rather than use '$\varphi(\mathsf{x})$' where we are recruiting the free variable 'x' for place-holding duties. But we will stick to the more common mathematical usage (even though Fregeans will sigh sadly). And a word to the wise: if you know what 'clash of variables' means, you will also know how we can avoid it in some future contexts by relabelling variables when necessary – so we won't fuss about that.

For a toy example, suppose the expressions of our theory's language L are built up from just seven basic symbols. Associate those with the digits 1 to 7, and associate the comma we might use to separate expressions in a sequence of expressions with the digit 8. Then a single L-expression, and also a sequence of L-expressions separated by commas, can be directly mapped to a sequence of digits, which can then be read as a single numeral in standard decimal notation, denoting a natural number. That mapping is the simplest of algorithms. And in reverse, undoing the coding is equally mechanical (though if the string of digits expressing some number contains the digit '9' or '0', the algorithm won't output any result when we try to decode it).

Which scheme of Gödel-numbering we adopt in practice will depend on considerations of ease of manipulation. In theory it won't matter: any effective scheme is as good as any other (as we will be able to effectively map codes for wffs or sequences of wffs produced by one scheme to codes produced by another, simply by decoding according to the first scheme and re-coding using the second).

3.4   Three new numerical properties/relations

Defn. 15. *Take an effectively axiomatized formal theory T, and fix on a scheme for Gödel-numbering expressions and sequences of expressions from T's language. Then, relative to that numbering scheme, we can define the following properties/relations:*

> *Wff(n) iff n is the Gödel number of a T-wff.*
> *Sent(n) iff n is the Gödel number of a T-sentence.*
> *Prf(m, n) iff m is the Gödel number of a T-proof of the T-sentence with code number n.*

So *Wff*, for example, is a numerical property which, so to speak, 'arithmetizes' the syntactic property of being a wff. Now, true enough, these three aren't the kind of numerical properties/relations you are familiar with. But they are perfectly well-defined. Indeed, we can say more:

Theorem 3. *Suppose T is an effectively axiomatized formal theory, and suppose we are given a Gödel-numbering scheme. Then the corresponding numerical properties/relations* Wff, Sent, Prf *are effectively decidable.*

Proof.[2] Take *Wff*. The number n has this property if and only if (i) n decodes into a string of T-symbols (an effective procedure a computer could carry out), and (ii) that string of T-symbols is a wff (which, since T has an effectively formalized language by assumption, again a computer could decide). In short, it is effectively decidable whether *Wff*(n).

The case of *Sent* is similar. And as for *Prf*, since T is an effectively axiomatized theory it is effectively decidable whether a supposed proof-array of the theory is

---

[2] Our end-of-proof symbol will be '⊠': we need the more usual '□' for other duties later.

the genuine article proving its purported conclusion. So it is effectively decidable whether the array, if any, which gets the code number m is actually a T-proof of the conclusion coded by n. That is to say, it is effectively decidable whether $Prf(m, n)$. ⊠

3.5 T can express Prf

So far, so straightforward. Now things get more exciting. In this section and the next, we state two key results, which will prepare the ground for our Gödelian proof of Theorem 1. For the moment, we will have to state the results without proof; later, we will see what it takes to prove (close variants) of them. But for now, we just want to explain what these two results claim. The first is as follows:

Theorem 4. *Suppose T is an effectively axiomatized theory which includes the language of basic arithmetic, and suppose we have fixed on a Gödel-numbering scheme. Then T can express the corresponding numerical relation Prf using some arithmetical wff* $\mathsf{Prf}(\mathsf{x}, \mathsf{y})$.

In other words, there is an arithmetic wff $\mathsf{Prf}(\mathsf{x}, \mathsf{y})$ such that $\mathsf{Prf}(\overline{\mathsf{m}}, \overline{\mathsf{n}})$ is true if and only if m codes for a T proof of the sentence with Gödel number n.

This theorem is not supposed to be obvious! How can we prove its perhaps surprising claim? We can take the low road. Take a particular T and trudge through the details of building a wff of basic arithmetic which indeed expresses the relation Prf for this theory T. Then we generalize, by noting that the same strategies and tricks that we use in the chosen particular case will apply equally when dealing with other effectively axiomatized theories.

Or we can take the high road. We start off by showing that the language of basic arithmetic is able to express decidable numerical properties and relations quite generally. And then we apply our sweeping result to the instances we are interested in: for we've just seen that the numerical relation Prf is decidable when T is a formalized theory. We will explore a version of this option in Chapter 10.

3.6   Defining a Gödel sentence G_T

With a predicate $\mathsf{Prf}(\mathsf{x}, \mathsf{y})$ available in the theory T to express the relation Prf, we can now add a further neat definition:

Defn. 16. *Put* $\mathsf{Prov}(\mathsf{x}) =_{\text{def}} \exists \mathsf{z} \mathsf{Prf}(\mathsf{z}, \mathsf{x})$ *(where the quantifier runs over all the numbers in the domain). Then* $\mathsf{Prov}(\overline{\mathsf{n}})$, *i.e.* $\exists \mathsf{z} \mathsf{Prf}(\mathsf{z}, \overline{\mathsf{n}})$, *is true iff some number Gödel-numbers a T-proof of the wff with Gödel-number n, i.e. is true just if the wff with code number n is a T-theorem. So* $\mathsf{Prov}(\mathsf{x})$ *is naturally called a* provability predicate.

And now comes the key result we need for building towards the First Theorem. Still working with the same theory T and Gödel-numbering scheme, and now for clarity's sake introducing subscripts to register which theory we are dealing with:

Theorem 5. *We can construct a Gödel sentence G_T for the theory T in the language of basic arithmetic with the following property: G_T is true if and only if $\neg\mathsf{Prov}_T(\overline{g})$ is true, where g is the code number of G_T.*

Don't worry for the moment about how we prove this theorem (it is surprisingly easy). Just note what our theorem implies. By construction, we said, G_T is true on interpretation iff $\neg\mathsf{Prov}_T(\overline{g})$ is true, i.e. iff the wff with Gödel number g is not a T-theorem, i.e. iff G_T is not a T-theorem. In short, G_T *is true if and only if it isn't a T-theorem.*

Stretching a point, it is rather as if G_T 'says' *I am unprovable in T.* (Of course, that *is* stretching a point: strictly speaking, G_T doesn't *really* say that – G_T is just a fancy sentence in the language of basic arithmetic, so it is in fact just about *numbers*, and doesn't refer to any wff. More about this later, in §11.2.) Still, with that point in mind, you will probably immediately spot that we can now prove ...

3.7 Incompleteness!

Here again is

Theorem 1. *Suppose T is a formal axiomatized theory whose language contains the language of basic arithmetic. Then, if T is sound, there will be a true sentence G_T of basic arithmetic such that $T \nvdash G_T$ and $T \nvdash \neg G_T$, so T is negation incomplete.*

Proof. Take G_T to be the Gödel sentence introduced in Theorem 5. Suppose T is sound and $T \vdash G_T$. Then G_T would be a theorem, and hence G_T – which is true iff it is not a T-theorem – would be false. So T would have a false theorem and hence T would not be sound, contrary to hypothesis. So $T \nvdash G_T$.

Hence G_T is not provable. Since it is true iff it is not provable, G_T is true after all. So $\neg G_T$ is false and T, being sound, can't prove that either. Therefore we also have $T \nvdash \neg G_T$.

So, in sum, T can't formally decide G_T one way or the other. T is negation incomplete. ⊠

This proof is very straightforward. So the devil is in the details of the proofs of the preliminary results we labelled as Theorems 4 and 5. As promised, later chapters will dig down to the relevant details.

Gödel's proof of the syntactic version of the incompleteness theorem, i.e. Theorem 2, also uses the same construction of a Gödel sentence, but this time we need to trade in the semantic assumption that T is sound for syntactic assumptions about what T can and can't prove. So we will need syntactic analogues of Theorems 4 and 5. Again more devilish detail. Again more about this in later chapters.

3.8   Gödel and the Liar

The claim is that, in a suitable theory T and using some Gödel coding, we can construct an arithmetic sentence G_T which holds if and only if G_T *is not provable in T*; and then such a sentence can neither be proved nor refuted in a sound T. But you might well be suspicious. After all, we know we fall into paradox if we try to construct a Liar sentence L which holds if and only if L *is not true*. So why does the construction of the Liar sentence lead to *paradox*, while the construction of the Gödel sentence give us a *theorem*?

Which is a very good question. You have exactly the right instincts in raising it. The coming chapters, however, aim to give you a convincing answer.

But we are touching here on the deep roots of the incompleteness theorem. Suppose T is an effectively axiomatized theory which can express enough arithmetic. Then, as we'll confirm later, T can express the property of being a provable T-sentence. But, as we will also confirm, T can't express the property of being a true T-sentence (if it could, then T would be beset by the Liar paradox). So the property of being a true T-sentence and the property of being a provable T-sentence must be different properties. Hence either there are true-but-unprovable-in-T sentences or there are false-but-provable-in-T sentences. Assuming that T is sound rules out the second option. So the truths of T's language outstrip T's theorems. Therefore T can't be negation complete. *That* might be said to be the Master Argument for incompleteness: see §16.5.

4 Undecidability and incompleteness

In Chapter 1, we introduced the idea of a negation-complete, effectively axiomatized, formal theory T.

We noted that if we are aiming to construct a theory of basic arithmetic, we would ideally like the theory to be able to prove *all* the truths expressible in the language of basic arithmetic, and hence to be negation complete (at least as far as statements of basic arithmetic are concerned). But Gödel's First Incompleteness Theorem tells us that that's impossible: roughly, a nice enough theory T will always be negation incomplete for basic arithmetic.

Now, as we noted in Chapter 2, the Theorem comes in two flavours, depending on whether we cash out the idea of being 'nice enough' in terms of (i) the semantic idea of T's being a *sound theory which uses enough of the language of arithmetic*, or (ii) the idea of T's being a *consistent theory which proves enough arithmetic*. Then we saw in Chapter 3 that Gödel's own proofs, of either flavour, go via the idea of numerically coding up inside arithmetic itself syntactic facts about what can be proved in T, and then constructing an arithmetical sentence that – in virtue of the coding – is true if and only if it is not provable (it is rather as if it says *I am not provable in T*).

We ended by remarking that, at least at the level of arm-waving description in Chapter 3, the Gödelian construction might look a bit worrying, with its echoes of the Liar Paradox. It might well go some way towards calming the worry that an illegitimate trick is being pulled if we now give a somewhat different proof of incompleteness. This proof will introduce the idea of a *diagonalization argument*. And as we will see later, it is diagonalization which is really the key to Gödel's own proof.

So now read on ...

4.1 Negation completeness and decidability

Let's start with another definition:

Defn. 17. *A theory T is* decidable *iff the property of being a theorem of T is an effectively decidable property – i.e. iff there is a mechanical procedure for determining, for any given sentence φ of T's language, whether $T \vdash \varphi$.*

Terminology check: a theory T formally *decides* a particular sentence φ iff either

$T \vdash \varphi$ or $T \vdash \neg\varphi$; a theory T is *decidable* iff for *any* sentence φ of its language we can effectively determine whether $T \vdash \varphi$. Two different notions then with similar terminology: in practice, though, you shouldn't get confused![1]

We can now easily show:

Theorem 6. *Any consistent, negation-complete, effectively axiomatized formal theory is decidable.*

Proof   For convenience, we can assume our theory T's proof system is a Frege/Hilbert axiomatic logic, where proofs are just linear sequences of wffs (but it should be pretty obvious how to generalize the argument to other kinds of proof systems, where proof arrays are arranged e.g. as trees of some kind).

Recall, we stipulated (in Defns. 2, 3) that if T is a properly formalized theory, its formalized language L has a finite number of basic symbols. Now, we can evidently put those basic symbols in some kind of 'alphabetical order', and then start mechanically listing off all the possible strings of symbols in order – e.g. the one-symbol strings, followed by the finite number of two-symbol strings in 'dictionary' order, followed by the finite number of three-symbol strings in 'dictionary' order, followed by the four-symbol strings, etc., etc.

Now, as we go along, generating strings of symbols, it will be a mechanical matter to decide whether a particular string is in fact a sequence of wffs. And if it is, it will be a mechanical matter to decide whether the sequence of wffs is a T-proof, i.e. to check whether each wff is either an axiom or follows from earlier wffs in the sequence by one of T's rules of inference. (That's all effectively decidable in a properly formalized theory, by Defns. 2, 3). If the sequence is a kosher well-constructed proof, finishing with a sentence φ, then list this wff φ as a T-theorem.

We can in this way start mechanically generating a list that will eventually contain any T-theorem (since any T-theorem is the last sentence of a proof).

And that enables us to decide, of an arbitrary sentence φ of our consistent, negation-complete T, whether it is indeed a T-theorem. Just start listing all the T-theorems. Since T is negation complete, eventually either φ or $\neg\varphi$ turns up (and then you can stop!). If φ turns up, declare it to be a theorem. If $\neg\varphi$ turns up, then since T is consistent, we can declare that φ is *not* a theorem.

Hence, there *is* a dumbly mechanical 'wait and see' procedure for deciding whether φ is a T-theorem, a procedure which (given our assumptions about T) is guaranteed to deliver a verdict in a finite number of steps.   ☒

We are, of course, relying here on a *very* relaxed notion of effective decidability-in-principle, where we aren't working under any practical time constraints or constraints on available memory etc. (so note, 'effective' doesn't mean 'practically efficacious' or 'efficient'). We might have to twiddle our thumbs for an

---

[1]To fix ideas, note that a theory can be decidable without deciding every wff. For example, the toy propositional theory T of §1.3 is decidable (as is familiar, because propositional logic is complete, a truth-table test will determine whether $T \vdash \varphi$ for any given wff φ of T's language). In particular, we can see that $T \nvdash$ q and $T \nvdash \neg$q. Therefore T doesn't decide q, so T doesn't decide every wff.

immense time before one of φ or $\neg\varphi$ turns up. Still, our 'wait and see' method is guaranteed in this case to produce a result in finite time, in an entirely mechanical way – so this counts as an effectively computable decision procedure in our official generous sense (see again the comments on Defn. 1).

4.2   Capturing numerical properties in a theory

Here's an equivalent way of rewriting part of an earlier definition:

Defn. 13.   *A numerical property P is expressed by the open wff $\varphi(\mathsf{x})$ with one free variable in a language L which contains the language of basic arithmetic iff, for every n,*
    i. if n has the property P, then $\varphi(\overline{n})$ is true,
    ii. if n does not have the property P, then $\neg\varphi(\overline{n})$ is true.

(Recall, $\overline{n}$ indicates L's standard numeral for n.) And now we want a new companion definition. Assume the language of T contains standard numerals: then

Defn. 18.   *The theory T captures the numerical property P by the open wff $\varphi(\mathsf{x})$ iff, for any n,*
    i. if n has the property P, then $T \vdash \varphi(\overline{n})$,
    ii. if n does not have the property P, then $T \vdash \neg\varphi(\overline{n})$.

Note the contrast: what a theory can *express* depends on the richness of its language (the definition doesn't mention proofs or theorems); what a theory can *capture* – mnemonic: <u>ca</u>se-by-case <u>p</u>rove – depends on what theorems can be derived in the theory.[2]

Just as a theory can express two-place relations (say) as well as monadic properties, a theory can capture relations as well as properties. So (for future reference) we expand our definition in the obvious way like this:

Defn 18. *(continued)*   *The theory T captures the two-place numerical relation R by the open wff $\varphi(\mathsf{x},\mathsf{y})$ iff, for any m, n,*
    i. if m has the relation R to n, then $T \vdash \varphi(\overline{m},\overline{n})$,
    ii. if m does not have the relation R to n, then $T \vdash \neg\varphi(\overline{m},\overline{n})$.

But for the moment, let's concentrate on the case of capturing properties.

Ideally, of course, we will want any competent theory of arithmetic not just to express but also to capture lots of numerical properties, i.e. to be able to prove particular numbers have or lack these properties. But what kinds of properties do we want to capture?

Well, suppose that P is some effectively decidable property of numbers, i.e. one for which there is a mechanical procedure for deciding, given a natural number n,

---

[2]To be honest, 'represents' is much more commonly used than my 'captures', but I'll stick here to the slightly idiosyncratic but memorable jargon adopted in *IGT2*. Terminology here is a mess: for example, some still use Kleene's 'numeralwise express' to mean (not our 'express' but) 'captures/represents'.

whether n has property P or not (see Defn. 1 again). So we can, in principle, run the procedure to decide whether n has this property P. Now, when we construct a formal theory of the arithmetic of the natural numbers, we will surely want deductions inside our theory to be able to track, case by case, any mechanical calculation that we can already perform informally (we see some examples of this in the next chapter). We don't want going formal to *diminish* our ability to determine whether n has a property P. Formalization aims at regimenting what we can in principle already do: it isn't supposed to hobble our efforts. So while we might have some passing interest in more limited theories, we will ideally aim for a formal theory T which at least (a) is able to frame some open wff $\varphi(\mathsf{x})$ which expresses the decidable property P, and (b) is such that if n has property P, $T \vdash \varphi(\bar{\mathsf{n}})$, and if n does not have property P, $T \vdash \neg\varphi(\bar{\mathsf{n}})$.

In short, we will want T not only to be able to express the decidable numerical property P but also to capture P in the sense of our definition. Focusing on the syntactic side of this, let's say:

Defn. 19. *A formal theory T is sufficiently strong iff it captures all effectively decidable numerical properties.*[3]

Then, in summary, it seems reasonable to want a formal theory of arithmetic to be sufficiently strong. When *we* can (or at least, given world enough and time, *could*) decide whether a particular number has a certain property, the *theory* should be able to do it.

4.3   Sufficiently strong theories are undecidable

We now prove a lovely theorem (take it slowly, savour it!):

Theorem 7. *No consistent, effectively axiomatized and sufficiently strong formal theory is decidable.*

Proof   We suppose T is a consistent and sufficiently strong theory yet also decidable, and derive a contradiction.

If T is sufficiently strong, it must have a supply of open wffs (for capturing numerical properties). And by Defn 2, it must be decidable what strings of symbols are T-wffs with the free variable 'x'. And we can use the dodge in the proof of Theorem 6 to start mechanically listing such wffs

$$\varphi_0(\mathsf{x}), \varphi_1(\mathsf{x}), \varphi_2(\mathsf{x}), \varphi_3(\mathsf{x}), \ldots.$$

For we can just start churning out all the strings of symbols of T's language (by length and in 'alphabetical order'), and as we go along we mechanically select out the wffs with free variable 'x'.

So now we can introduce the following definition:

---

[3]It would be equally natural, of course, to require that the theory also capture all decidable relations and all computable functions – but for present purposes we don't need to worry about that.

n has the property D if and only if $T \vdash \neg\varphi_n(\overline{n})$.

That's a perfectly coherent stipulation. Of course, property D isn't presented in the familiar way in which we ordinarily present properties of numbers: but our definition tells us what has to be the case for n to have the property D, and that's all we will need.

Now for the key observation: our supposition that T is a decidable theory entails that D is an effectively decidable property of numbers.

Why? Well, given any number n, it will be a mechanical matter to start listing off the open wffs until we get to the n-th one, $\varphi_n(\mathsf{x})$. Then it is a mechanical matter to form the numeral $\overline{n}$, substitute it for the variable, and then prefix a negation sign. Now we just apply the supposed mechanical procedure for deciding whether a sentence is a T-theorem to test whether the resulting wff $\neg\varphi_n(\overline{n})$ is a theorem. So, on our current assumptions, there is an algorithm for deciding whether n has the property D.

Since, by hypothesis, the theory T is sufficiently strong, it can capture all decidable numerical properties. So it follows, in particular, that D is capturable by some open wff. This wff must of course eventually occur somewhere in our list of the $\varphi(\mathsf{x})$. Let's suppose the d-th wff does the trick: that is to say, property D is captured by $\varphi_d(\mathsf{x})$.

It is now entirely routine to get out a contradiction. For, just by the definition of capturing, to say that $\varphi_d(\mathsf{x})$ captures D means that for any n,

> if n has the property D, $T \vdash \varphi_d(\overline{n})$,
> if n doesn't have the property D, $T \vdash \neg\varphi_d(\overline{n})$.

So taking in particular the case $n = d$, we have

> i. if d has the property D, $T \vdash \varphi_d(\overline{d})$,
> ii. if d doesn't have the property D, $T \vdash \neg\varphi_d(\overline{d})$.

But note what our initial definition of the property D above implies for the particular case $n = d$:

> iii. d has the property D if and only if $T \vdash \neg\varphi_d(\overline{d})$.

From (ii) and (iii), it follows that whether d has property D or not, the wff $\neg\varphi_d(\overline{d})$ is a theorem either way. So by (iii) again, d does have property D, hence by (i) the wff $\varphi_d(\overline{d})$ must be a theorem too. So a wff and its negation are both theorems of T. Which makes T inconsistent.

In sum, the supposition that T is a consistent and sufficiently strong axiomatized formal theory *and* is decidable leads to contradiction. $\boxtimes$

We have shown then that if T is properly formalized, consistent and can prove enough arithmetic, then there is no way of mechanically determining what's a T-theorem and what isn't.

4.4 Diagonalization

Let's highlight the key construction here. In defining the property D, for each n, we take the n-th wff $\varphi_n(\mathsf{x})$, and plug in the standard numeral for the index n (before taking the negation of the result). This sort of thing is called *diagonalization*. Why?

Well, just imagine the square array you get by writing $\varphi_0(\overline{0})$, $\varphi_0(\overline{1})$, $\varphi_0(\overline{2})$, etc. in the first row, $\varphi_1(\overline{0})$, $\varphi_1(\overline{1})$, $\varphi_1(\overline{2})$, etc. in the next row, $\varphi_2(\overline{0})$, $\varphi_2(\overline{1})$, $\varphi_2(\overline{2})$ etc. in the next row, and so on. *Then the wffs of the form $\varphi_n(\overline{n})$, including $\varphi_d(\overline{d})$, lie down the diagonal through the array.*

We'll be meeting other instances of this sort of construction. And it is a diagonalization of this kind that is really at the heart of Gödel's incompleteness proof.[4]

4.5 Incompleteness again!

So we have now shown:

Theorem 6. *Any consistent, negation-complete, effectively axiomatized formal theory is decidable.*

Theorem 7. *No consistent, effectively axiomatized and sufficiently strong formal theory is decidable.*

We can therefore deduce:

Theorem 8. *A consistent, effectively axiomatized, sufficiently strong, formal theory cannot be negation complete.*

Wonderful! A seemingly remarkable theorem, proved remarkably quickly (this time without having to simply assume unproved lemmas along the way).[5]

Note, though, that – unlike Gödel's own proof strategy – Theorem 8 doesn't actually yield a specific undecidable sentence for a given theory T. And more importantly, the interest of the theorem depends on the still-informal notion of a 'sufficiently strong' theory being in good order. Have we perhaps just shown that looking for sufficient strength is, after all, an unreasonable demand?

Now, I wouldn't have written up the argument in this chapter if this notion of T's being sufficiently strong were intrinsically problematic. Still, we are left with a major task here: we will need to give a sharper account of what makes for an effectively decidable property in order to (i) clarify the notion of sufficient strength, while (ii) making it plausible that we really do want theories to be sufficiently strong in this clarified sense.

---

[4]The grandfather of all such uses of diagonalization is Cantor's diagonal argument (see e.g. the Wikipedia entry, as well as *IGT2*, §2.5).

[5]I learnt the argument for Theorem 8 as a student – so decades ago! – from lectures by Timothy Smiley.

This can be done. However, supplying and defending the needed sharp account of the notion of effective decidability takes quite a bit of work. And we don't need to do the work in order to prove core versions of the First Incompleteness Theorem via Gödel's original method as partially sketched in Chapter 3. So over the coming chapters, we are going to revert to exploring something closer to Gödel's route to the incompleteness theorems.

5 Two weak arithmetics

So far we have talked rather abstractly of theories which 'can prove a certain modest amount of arithmetic' and about theories which are 'sufficiently strong'. But we haven't said what such theories look like. In fact, we haven't yet looked at *any* detailed theory of arithmetic. It is obviously high time that we stopped operating at the level of abstraction of earlier chapters; we need to start getting our hands dirty.

This chapter, then, introduces a couple of weak arithmetics ('arithmetics', that is to say, in the sense of 'theories of arithmetic'). We first meet Baby Arithmetic (as a warm-up) and then the important Robinson Arithmetic. You can by all means skip lightly over the more boring proof details here; but you do need to get a sense of how these two weaker formal theories work, in preparation for the next chapter where we introduce the much stronger Peano Arithmetic.

5.1 The language L_B

First we describe *the language of baby arithmetic*, L_B. Its symbols, with their built-in interpretations, are

0	proper name denoting zero
S, +, ×	function symbols for, respectively, the successor, addition and multiplication functions
(,)	parentheses for use with +, ×
=, ¬	logical vocabulary, just the identity predicate and negation.

We write the one-place successor function symbol in 'prefix' position, so we can form the standard numerals 0, S0, SS0, SSS0, ... (see §1.5). Recall, we use '$\overline{n}$' to represent the standard numeral SS...S0 with n occurrences of 'S'. Thus '$\overline{3}$' is short for 'SSS0' which denotes the number 3.

We will however write '+' and '×' as 'infix' function symbols in the usual way – i.e. we write (S0 + SS0) rather than prefix the function sign as in +S0SS0. So we need the parentheses for scoping the function signs, to disambiguate S0 + SS0 × SSS0, e.g. as (S0 + (SS0 × SSS0)). For readability, though, we will follow common practice and usually drop outermost pairs of brackets.

From these symbols, we can construct the *terms* of L_B. A term is a referring expression built up from occurrences of '0' and applications of the function

28

expressions 'S', '+', '×'. Examples are 0, SSS0, S0 + SS0, (S0 + SS0) × SSS0, SSS0 + ((S0 + SS0) × SSS0), and so on.

We will use σ and τ as metalinguistic placeholders for terms of L_B. The *value* of a term τ is the number it denotes when standardly interpreted: the values of our example terms are respectively 0, 3, 3, 9 and 12.

The sole built-in predicate of the language L_B is the logical identity sign. Since L_B lacks other non-logical predicates, the only way of forming atomic wffs in the language is therefore by taking two terms and putting the identity sign between them. In other words, the atomic wffs of L_B are *equations* relating terms denoting particular numbers. So, for example, S0 + SS0 = SSS0 is a true atomic wff – which we can abbreviate, as $\overline{1} + \overline{2} = \overline{3}$. And S0 + SS0 = SS0 × SS0 is a false atomic wff – which we can abbreviate, as $\overline{1} + \overline{2} = \overline{2} \times \overline{2}$.

We now add a negation sign to the language L_B so that we can also explicitly assert that various equations do *not* hold. For example, ¬ S0 + SS0 = SS0 × SS0 is true. Though, for readability's sake, we will prefer to abbreviate that last wff as S0 + SS0 ≠ SS0 × SS0.[1]

5.2   The axioms and logic of Baby Arithmetic

The theory BA couched in this language L_B will come equipped with a classical deductive system to deal with negation and identity. You can choose your favourite system. In illustrations, we'll set out proofs in a Fitch-like natural deduction format (because it is likely to be familiar, and is easy to follow even if it isn't): but absolutely nothing hangs on the choice.

Next, we want non-logical axioms governing the successor function. We want to capture the idea that, if we start from zero and repeatedly apply the successor function, we keep on getting further numbers – i.e. different numbers have different successors: contraposing, for any m, n, if $Sm = Sn$ then $m = n$. Further, zero isn't a successor, i.e. we never cycle back to zero: for any n, $0 \neq Sn$.

However, there are no quantifiers in L_B. So we can't directly express those general facts about the successor function inside the object language L_B. Rather, we have to employ *schemas* (i.e. general templates) and use the generalizing apparatus in our English metalanguage to say: *any sentence that you get from one of the following schemas by substituting standard numerals for the place-holders 'ζ', 'ξ' is an axiom.*

Schema 1. $0 \neq S\zeta$

Schema 2. $S\zeta = S\xi \rightarrow \zeta = \xi$

NB: These schemas aren't axioms of BA; the Greek metavariables don't belong to the language L_B. It is, to repeat, *instances* of the schemas got by systematically

---

[1] Fine print: we can allow an equation to be preceded by any number of negation signs, with the result counting as a (non-atomic) wff. But we are going to give L_B a classical logic, so a pair of adjacent negation signs just cancel each other out. Hence we need only worry about atomic formulae and their negations. It wouldn't make a significant difference if we also gave L_B the other connectives. But it is crucial that L_B lacks the apparatus of quantification.

replacing the placeholders with numerals – same placeholder, same replacement – which are the axioms.[2] We'll see some examples in a moment.

Next, we want non-logical axioms for addition that capture some key ideas underlying the school-room rules. So first we claim that adding zero to a number makes no difference: for any m, $m + 0 = m$. And next, adding a non-zero number Sn (i.e. $n + 1$) to m is governed by the following rule: for any m, n, $m + Sn = S(m + n)$ – i.e. $m + (n + 1) = (m + n) + 1$. These two principles together tell us how to add zero to a given number m; and then adding one is defined as the successor of the result of adding zero; and then adding two is defined as the successor of the result of adding one; and so on up – thus defining adding n for any particular natural number n.

Note that because of L_B's lack of quantifiers, we again can't express all that directly inside L_B itself. We again have to resort to schemas, and say that anything you get by substituting standard numerals for placeholders in one of the following schemas is an axiom – for short, *every numeral instance of these schemas is an axiom*:

Schema 3. $\quad \zeta + 0 = \zeta$

Schema 4. $\quad \zeta + S\xi = S(\zeta + \xi)$

We can similarly pin down the multiplication function by requiring that *every numeral instance of these schemas too is an axiom*:

Schema 5. $\quad \zeta \times 0 = 0$

Schema 6. $\quad \zeta \times S\xi = (\zeta \times \xi) + \zeta$

Instances of Schema 5 tell us the result of multiplying by zero. Instances of Schema 6 with 'ξ' replaced by '0' define how to multiply by one in terms of first multiplying by zero and then applying the already-defined addition function. Once we know about multiplying by one, we can use another instance of Schema 6 – this time with 'ξ' replaced by '$S0$' – to tell us how to multiply by two (multiply by one and then do some addition). And so on, thus defining multiplication for every number.

To summarize, then,

Defn. 20. BA, *Baby Arithmetic, is the theory whose language is L_B, whose logic comprises classical negation rules and identity rules, and whose non-logical axioms are every numeral instance of Schemas (1) to (6).*

Now, suppose we had adopted versions of our everyday English numerals to denote numbers in a formal arithmetic. Then we would also need a whole bunch of additional axioms like S zero = one, S one = two, and so on, plus further school-room rules to deal with our base-ten notation. We have avoided this sort of

---

[2]Fine print: it wouldn't actually make any difference to the strength of our theory if we allowed the placeholding metavariables to be systematically replaced by any terms, not just standard numerals. But let's keep things simple.

complication by choosing to use our standard numerals to denote numbers in BA. So that's a non-trivial choice. We are already building into our system for denoting numbers something of the structure of number series, and it is this which enables our axioms for BA to be so simple. Which is all perfectly legitimate: but we should be aware that this *is* what we are doing.

Note, finally, that although BA's axioms fall into just six kinds, there are an infinite number of them – since *any* instance of our schemas counts as an axiom. However, although it isn't *finitely* axiomatized, it is still an *effectively* axiomatized theory: given a candidate wff, we can effectively decide whether it is an instance of one of those six schemas and hence an axiom.

5.3   Some proofs inside BA

Let's start with three brisk examples of how arithmetic can be done inside BA, as we laboriously break down some informal calculations into minimal steps.

First, let's show that BA $\vdash 0 + \overline{2} = \overline{2}$. In other words, $0 + SS0 = SS0$ is a theorem – and note carefully, this wff *isn't* an instance of Schema 3.

1.	$0 + 0 = 0$	Axiom, instance of Schema 3
2.	$0 + S0 = S(0 + 0)$	Axiom, instance of Schema 4
3.	$0 + S0 = S0$	From 1, 2 by the identity laws
4.	$0 + SS0 = S(0 + S0)$	Axiom, instance of Schema 4
5.	$0 + SS0 = SS0$	From 3, 4 by the identity laws

The relevant identity law here is of course *Leibniz's Law* – LL, for short – which allows us to substitute terms which are proven equal.

Similarly, we can prove $\overline{2} + \overline{2} = \overline{4}$, i.e. $SS0 + SS0 = SSSS0$:

1.	$SS0 + 0 = SS0$	Axiom, instance of Schema 3
2.	$SS0 + S0 = S(SS0 + 0)$	Axiom, instance of Schema 4
3.	$SS0 + S0 = SSS0$	From 1, 2 by LL
4.	$SS0 + SS0 = S(SS0 + S0)$	Axiom, instance of Schema 4
5.	$SS0 + SS0 = SSSS0$	From 3, 4 by LL

And now let's show that BA $\vdash \overline{2} \times \overline{2} = \overline{4}$. In unabbreviated form, we need (rather laboriously!) to derive $SS0 \times SS0 = SSSS0$:

1.	$SS0 \times 0 = 0$	Axiom, instance of Schema 5
2.	$SS0 \times S0 = (SS0 \times 0) + SS0$	Axiom, instance of Schema 6
3.	$SS0 \times S0 = 0 + SS0$	From 1, 2 by LL
4.	$0 + SS0 = SS0$	Derived as in first proof above
5.	$SS0 \times S0 = SS0$	From 3, 4 by LL
6.	$SS0 \times SS0 = (SS0 \times S0) + SS0$	Axiom, instance of Schema 6
7.	$SS0 \times SS0 = SS0 + SS0$	From 5, 6 by LL
8.	$SS0 + SS0 = SSSS0$	Derived as in second proof above
9.	$SS0 \times SS0 = SSSS0$	From 7, 8 by LL

OK: so now let's generalize. Suppose that for some other m we'd started instead from the Axiom $\overline{m} + 0 = \overline{m}$, another instance of Schema 3. Then by similar steps as for the first two proofs, we can derive $\overline{m} + SS0 = SS\overline{m}$, i.e. $\overline{m} + \overline{2} = \overline{m+2}$ (here, $\overline{m+2}$ of course stands in for the standard numeral for $m+2$).

And then, generalizing further, if we keep extending the same proof idea with a few more steps cut to the same pattern, we can get BA to show $\overline{m} + \overline{3} = \overline{m+3}$, and $\overline{m} + \overline{4} = \overline{m+4}$, and so on. In fact, for any m, n, $\mathsf{BA} \vdash \overline{m} + \overline{n} = \overline{m+n}$.

Similarly, looking at the second proof pattern, we see that we'll be able to similarly prove $\overline{m} \times \overline{2} = \overline{m \times 2}$ for any m. And then, generalizing further, if we keep extending the same proof idea with more steps cut to the same pattern, we can prove $\overline{m} \times \overline{3} = \overline{m \times 3}$, and $\overline{m} \times \overline{4} = \overline{m \times 4}$, and so on. Take any m, n: then $\mathsf{BA} \vdash \overline{m} \times \overline{n} = \overline{m \times n}$.

We can now generalize a step further: BA can correctly evaluate not just the simplest terms but *all* terms of its language. That is to say,

Theorem 9. *Suppose τ is a term of L_B and suppose the value of τ on the intended interpretation of the symbols is t. Then $\mathsf{BA} \vdash \tau = \overline{t}$.*

Why so? Well, let's take a very simple example and then draw a general moral. Suppose we want to show e.g. that $(\overline{2} + \overline{3}) \times (\overline{2} \times \overline{2}) = \overline{20}$ – you'll forgive me for not writing out '$\overline{20}$' in basic notation with its twenty occurrences of 'S'! Then we can proceed as follows.

1. $(\overline{2} + \overline{3}) \times (\overline{2} \times \overline{2}) = (\overline{2} + \overline{3}) \times (\overline{2} \times \overline{2})$ Identity law
2. $\overline{2} + \overline{3} = \overline{5}$ BA can do simple addition
3. $(\overline{2} + \overline{3}) \times (\overline{2} \times \overline{2}) = \overline{5} \times (\overline{2} \times \overline{2})$ From 1, 2 by LL
4. $\overline{2} \times \overline{2} = \overline{4}$ BA can do simple multiplication
5. $(\overline{2} + \overline{3}) \times (\overline{2} \times \overline{2}) = \overline{5} \times \overline{4}$ From 3, 4 by LL
6. $\overline{5} \times \overline{4} = \overline{20}$ BA can do simple multiplication
7. $(\overline{2} + \overline{3}) \times (\overline{2} \times \overline{2}) = \overline{20}$ From 5, 6 using LL

What we do here is 'evaluate' the complex formula on the right 'from the inside out', reducing the complexity of what's on the right at each stage, and hence eventually equating the complex formula on the left with a standard numeral on the right. Evidently, we can always do this trick, whatever complex formula we start from.

Next, we note that BA knows that different standard numerals are indeed not equal. For example, let's show that $\mathsf{BA} \vdash \overline{4} \neq \overline{2}$.

1. $SSSS0 = SS0$ Supposition
2. $SSSS0 = SS0 \rightarrow SSS0 = S0$ Axiom, instance of Schema 2
3. $SSS0 = S0$ From 1, 2 by Modus Ponens
4. $SSS0 = S0 \rightarrow SS0 = 0$ Axiom, instance of Schema 2
5. $SS0 = 0$ From 3, 4 by Modus Ponens
6. $0 \neq SS0$ Axiom, instance of Schema 1
7.   Contradiction!   From 5, 6 and identity rules
8. $SSSS0 \neq SS0$ From 1 to 7, by Reductio ad Absurdum.

And a little reflection on this illustrative proof should now convince you that:

Theorem 10. *If s and t are distinct numbers, then $\mathsf{BA} \vdash \overline{s} \neq \overline{t}$.*

5.4 BA is a sound and negation-complete theory of the truths of L_B

We can conclude from our last two theorems that

Theorem 11. *For any L_B wff φ, if φ is true then* BA $\vdash \varphi$, *and if φ is false then* BA $\vdash \neg\varphi$. *In a phrase,* BA *correctly decides every L_B wff.*

Proof. The only wffs of BA are equations preceded by zero or more negation signs. Since our logic is classical, we can ignore pairs of negation signs, so any wff φ is equivalent to either (a) $\sigma = \tau$ or (b) $\sigma \neq \tau$, for some terms σ, τ. Let σ evaluate to s and τ evaluate to t. Then by Theorem 9, (i) BA $\vdash \sigma = \bar{s}$ and (ii) BA $\vdash \tau = \bar{t}$.

And now we just consider the four possible cases.

(a) Suppose φ is equivalent to $\sigma = \tau$, and is true because $s = t$. Then $\bar{s}$ must be the very same numeral as $\bar{t}$. We can therefore immediately conclude from (i) and (ii) that BA $\vdash \sigma = \tau$ by the logic of identity. So BA $\vdash \varphi$.

(b) Suppose alternatively that φ is equivalent to $\sigma = \tau$, and is false because $s \neq t$. Then by Theorem 10, BA $\vdash \bar{s} \neq \bar{t}$, and together with (i) and (ii) that implies BA $\vdash \sigma \neq \tau$, again by the logic of identity. So BA $\vdash \neg\varphi$.

(c) Suppose φ is equivalent to $\sigma \neq \tau$, and is true because $s \neq t$. Then as in (b), BA $\vdash \sigma \neq \tau$. So now BA $\vdash \varphi$.

(d) Suppose finally that φ is equivalent to $\sigma \neq \tau$ and is false because $s = t$. As in (a), BA $\vdash \sigma = \tau$, hence BA $\vdash \neg\varphi$.

Hence, BA correctly decides every L_B wff φ. ⊠

We therefore immediately have

Theorem 12. BA *is a sound effectively axiomatized theory which is negation complete.*

Proof. BA is evidently a sound theory – all its axioms are trivial arithmetical truths, and its logic is truth-preserving, so all its theorems are true. It is effectively axiomatized. And we've just seen that for every wff φ, it proves the true one of φ and $\neg\varphi$; and so is a negation complete theory. ⊠

"Hold on! I thought we couldn't have a sound effectively axiomatized theory of arithmetic which is negation complete." No. Theorem 1 didn't say *that*: it said we couldn't have a sound, negation-complete, effectively axiomatized theory which contains what we called the language of basic arithmetic – and *that* language allows us to quantify over numbers. By contrast, L_B is quantifier-free. This language only allows us to express facts about adding and multiplying particular numbers (it can't express numerical generalizations). That's why it can be complete.

"Ah. So having quantifiers in a theory's language can make all the difference?" Yes!

5.5   Robinson Arithmetic, Q

So far that is all very straightforward, but also rather unexciting.[3] The reason that Baby Arithmetic manages to prove every correct claim that it can express – and is therefore negation complete by our Defn. 8 – is that it can't express very much. In particular, as we just stressed, it can't express any generalizations at all. And so the obvious way to start beefing up BA into something more expressively competent is to restore the familiar apparatus of quantifiers and variables. That's what we'll start doing.

First, then, we define the interpreted *first-order language of basic arithmetic* L_A (compare §1.5). We will keep the same non-logical vocabulary as in L_B: so there is still just a single non-logical constant denoting zero, plus the three function-symbols, S, +, ×, still expressing successor, addition and multiplication. But now we allow ourselves the full linguistic resources of first-order logic; so we now have all the connectives plus the usual supply of quantifiers and variables to express generality, as well as the built-in identity predicate. We fix the domain of the quantifiers to be the natural numbers. The result is the language L_A: this is the least ambitious language which 'contains the language of basic arithmetic' in the sense of Defn. 11.

Now for *Robinson Arithmetic*, commonly denoted simply 'Q'. This is a theory built in the formal language L_A, and coming equipped with a full first order proof system for classical logic. And for its non-logical axioms, now that we have the quantifiers available to express generality, we can replace each of BA's metalinguistic schemas (specifying an infinite number of formal axioms governing particular numbers) by a single generalized Axiom expressed inside L_A itself. For example, we can replace the first two schemas governing the successor function by the following:

Axiom 1. $\forall x(0 \neq Sx)$

Axiom 2. $\forall x \forall y(Sx = Sy \rightarrow x = y)$

Obviously, each instance of our earlier Schemas 1 and 2 can be deduced from the corresponding Axiom by instantiating the quantifiers with numerals.

These Axioms tell us that zero isn't a successor, but they don't explicitly rule out there being other objects that aren't successors cluttering up the domain of quantification. We didn't need to fuss about this before, because by construction BA can only talk about the numbers represented by standard numerals in the sequence '0, S0, SS0, . . .'. But now we have the quantifiers in play. And these

---

[3]Mathematically unexciting, anyway. But there is perhaps some philosophical interest. For we might reasonably suppose that the axiom schemas of BA at least partially encapsulate the meanings of the symbols for the symbol for zero and for the successor, addition and multiplication functions – they partially define what we are talking about. So it is consequently quite tempting to be a logicist at least about the arithmetic truths proved by BA, regarding them as truths of logic-plus-definitions. And this success might encourage us to pursue some ambitious form of logicism (see §1.5).

quantifiers are intended to run over the natural numbers; we certainly don't intend them to be also running over stray objects that aren't successors and aren't zero either.

So let's reflect that intention in an axiom which says that, other than zero, every number is a successor:

Axiom 3. $\forall x(x \neq 0 \rightarrow \exists y(x = Sy))$

Next, we can similarly replace our previous schemas for addition and multiplication by universally quantified Axioms in the obvious way:

Axiom 4. $\forall x(x + 0 = x)$

Axiom 5. $\forall x \forall y(x + Sy = S(x + y))$

Axiom 6. $\forall x(x \times 0 = 0)$

Axiom 7. $\forall x \forall y(x \times Sy = (x \times y) + x)$

Again, each of these axioms entails all the instances of BA's corresponding schema. In sum, then:

Defn. 21. *The formal theory with language* L_A, *Axioms 1 to 7, plus a classical first-order logic, is standardly called* Robinson Arithmetic, *or simply* Q.

Since any BA axiom can be derived from one of our new Q Axioms, anything that can be proved in BA can be proved in Q.

5.6   Robinson Arithmetic is not complete

Like BA, Q too is an effectively axiomatized sound theory. Its axioms are all true; and its logic is truth-preserving; so its derivations are proper proofs in the intuitive sense of demonstrations of truth. Every theorem of Q is a true L_A wff, then. But just which truths of L_A are theorems of Q?

Well, on the positive side,

Theorem 13. Q *correctly decides every quantifier-free* L_A *sentence. In other words,* $Q \vdash \varphi$ *if the quantifier-free wff* φ *is true, and* $Q \vdash \neg\varphi$ *if the quantifier-free wff* φ *is false.*

Proof. We know that Q (like BA) will correctly decide every atomic wff, i.e. correctly decide every equation between terms. We can then appeal to a background result of propositional logic which tells us that Q must then correctly decide every wff built up from those atoms using just the truth-functional propositional connectives.[4]   ⊠

---

[4]If you insist on details: Suppose the wff φ contains the atoms $\alpha_1, \alpha_2, \ldots, \alpha_n$, and consider any assignment V of truth-values to those atoms. Let α_n^V be α_n if that is true on V, and $\neg\alpha_n$ otherwise. Similarly, let φ^V be φ if that is true on V, and $\neg\varphi$ otherwise. So consider the line of a truth table for φ corresponding to the particular valuation V. On the left of the

So far, so good. However, there are very simple true *quantified* sentences that Q can't prove.

For example, while Q can prove any particular wff of the form $0 + \bar{n} = \bar{n}$, *it can't prove the corresponding universal generalization*:

Theorem 14. $Q \nvdash \forall x (0 + x = x)$.

Proof Since Q is a theory with a standard first-order theory, for any L_A-sentence φ, $Q \vdash \varphi$ only if $Q \vDash \varphi$ (that's just the soundness theorem for first-order logic). Hence one way of showing that $Q \nvdash \varphi$ is to show that $Q \nvDash \varphi$: and we can show *that* by producing a countermodel to the entailment – i.e. by finding an interpretation (a deviant, unintended, 'non-standard', re-interpretation) for L_A's wffs which makes Q's axioms true-on-that-interpretation but which makes φ false.

So here goes: take the domain of our deviant, unintended, re-interpretation to be the set N^* which comprises the natural numbers but with two other 'rogue' elements a and b added (these could be e.g. Kurt Gödel and his friend Albert Einstein – but any other pair of distinct non-numbers will do). Let '0' still refer to zero. And take 'S' now to pick out the successor* function S^* which is defined as follows: $S^* n = Sn$ for any natural number in the domain, while for our rogue elements $S^* a = a$, and $S^* b = b$. It is very easy to check that Axioms 1 to 3 are still true on this deviant interpretation. Zero is still not a successor. Different elements have different successors. And every non-zero element is again a successor (perhaps a self-successor! – though not necessarily an eventual successor of zero).

We now need to extend this interpretation to re-interpret the function-symbol '+'. Suppose we take this to pick out addition*, where $m +^* n = m + n$ for any natural numbers m, n in the domain, while $a +^* n = a$ and $b +^* n = b$. Further, for any x (whether number or rogue element), $x +^* a = b$ and $x +^* b = a$. If you prefer that in a matrix, read off *row $+^*$ column*:

$+^*$	n	a	b
m	$m + n$	b	a
a	a	b	a
b	b	b	a

---

table, this line assigns values to the atoms, and tells us that $\alpha_1^V, \alpha_2^V, \ldots, \alpha_n^V$ are all true. The corresponding assignment of a value to φ tells us that φ^V true. And the background result from propositional logic that we need is that $\alpha_1^V, \alpha_2^V, \ldots, \alpha_n^V \vdash_{PL} \varphi^V$.

Roughly, then, propositional logic can prove what's said by the line of a truth table for φ corresponding to the valuation V. (You might possibly recognize this as the result we need when proving the completeness of propositional logic by Kalmár's method.)

Now, a quantifier-free L_A sentence φ is built up using propositional connectives from atoms α of the form $\sigma = \tau$. Consider the valuation V that assigns these atoms their actual values. As in the proof of Theorem 11, Q proves the true atoms and proves the negations of the false ones, so Q proves each α^V. Hence by the background result it also proves φ^V, i.e. proves whichever is the true one of φ and $\neg \varphi$.

It is again easily checked that interpreting '+' as addition* still makes Axioms 4 and 5 true.[5]

We are not quite done, however, as we still need to show that we can give a co-ordinate re-interpretation of '×' in Q by some deviant multiplication* function. We can leave it as an exercise to fill in suitable details. Then, with the details filled in, we will have an overall interpretation which makes the axioms of Q true and $\forall x(0 + x = x)$ false. So $Q \nvdash \forall x(0 + x = x)$ ⊠

Theorem 15. Q *is negation incomplete.*

Proof. Put $\varphi = \forall x(0 + x = x)$. We've just shown that $Q \nvdash \varphi$. But obviously, Q can't prove $\neg\varphi$ either. Revert to the standard interpretation built into L_A. All Q's theorems are true but $\neg\varphi$ is false on that interpretation. So $\neg\varphi$ can't be a theorem. Hence φ is formally undecidable in Q.   ⊠

Of course, we've already announced that Gödel's incompleteness theorem is going to prove that *no* sound axiomatized theory whose language is at least as rich as L_A can be negation complete – that was Theorem 1. But we don't need to invoke anything as elaborate as Gödel's arguments to see that Q is incomplete. Q is, so to speak, *boringly* incomplete.

5.7   Statements of order in Robinson Arithmetic

Let's now start thinking about the properties and relations that can be captured in Robinson Arithmetic – recalling the definition of §4.2.

Here in particular is an example that will be useful:

Theorem 16. *In Robinson Arithmetic, the* less-than-or-equal-to *relation is not just expressed but captured by the wff* $\exists v(v + x = y)$.

It is obvious that the wff expresses the relation. So what we need to show is that, for any particular pair of numbers, m, n, if $m \leq n$, then $Q \vdash \exists v(v + \overline{m} = \overline{n})$, and if $m > n$, then $Q \vdash \neg\exists v(v + \overline{m} = \overline{n})$.

Proof. Suppose $m \leq n$, so for some $k \geq 0$, $k + m = n$. Q can prove everything BA proves and hence, in particular, can prove every true addition equation. So we have $Q \vdash \overline{k} + \overline{m} = \overline{n}$. But then $\exists v(v + \overline{m} = \overline{n})$ follows by existential quantifier introduction. Therefore $Q \vdash \exists v(v + \overline{m} = \overline{n})$, as was to be shown.

Suppose alternatively $m > n$. We need to show $Q \vdash \neg\exists v(v + \overline{m} = \overline{n})$. We'll first demonstrate this in the case where $m = 2$, $n = 1$, using a Fitch-style proof system. For brevity we will omit statements of Q's axioms and some other trivial steps; we drop unnecessary brackets too.

---

[5]In headline terms: For Axiom 4, we note that adding* zero on the right always has no effect. For Axiom 5, just consider cases. (i) $m +^* S^*n = m + Sn = S(m + n) = S^*(m +^* n)$ for 'ordinary' numbers m, n in the domain. (ii) $a + S^*n = a = S^*a = S^*(a +^* n)$, for 'ordinary' n. Likewise, (iii) $b + S^*n = S^*(b +^* n)$. (iv) $x +^* S^*a = x + a = b = S^*b = S^*(x +^* a)$, for any x in the domain. (v) Finally, $x +^* S^*b = S^*(x +^* b)$. Which covers every possibility.

1.	$\exists v(v + SS0 = S0)$	Supposition
2.	$a + SS0 = S0$	Supposition
3.	$a + SS0 = S(a + S0)$	From Axiom 5
4.	$S(a + S0) = S0$	From 2, 3 by LL
5.	$a + S0 = S(a + 0)$	From Axiom 5
6.	$SS(a + 0) = S0$	From 4, 5 by LL
7.	$a + 0 = a$	From Axiom 4
8.	$SSa = S0$	From 6, 7 by LL
9.	$SSa = S0 \rightarrow Sa = 0$	From Axiom 2
10.	$Sa = 0$	From 8, 9 by Modus Ponens
11.	$0 = Sa$	From 10
12.	$0 \neq Sa$	From Axiom 1
13.	Contradiction!	From 11, 12
14.	Contradiction!	$\exists$E 1, 2–13
15.	$\neg\exists v(v + SS0 = S0)$	From 1–14 by Reductio

The only step to explain may be at line (14) where we use a version of the Existential Elimination rule: if the temporary supposition $\varphi(a)$ leads to contradiction, for arbitrary a, then $\exists v\varphi(v)$ must lead to contradiction.

And having constructed a proof for the case $m = 2$, $n = 1$, inspection reveals that we can use the same general pattern of argument to show $Q \vdash \neg\exists v(v + \overline{m} = \overline{n})$ whenever $m > n$. So we are done. $\boxtimes$

Given the theorem we have just proved, we can sensibly add the standard symbol '$\leq$' to L_A, the language of Q, defined so that whatever terms – not just numerals – we put for the placeholders 'ξ' and 'ζ', $\xi \leq \zeta$ is just short for $\exists v(v + \xi = \zeta)$.[6] And then Q will be able to prove at least the expected facts about the less-than-or-equals relations among quantifier-free terms.

Note, by the way, that some presentations treat '$\leq$' as a primitive symbol built into our formal theories like Q from the start, governed by its own additional axiom(s). But nothing important hangs on the difference between that approach and our policy of introducing the symbol by definition. And of course, nothing hangs either on our policy of introducing '$\leq$' as our basic symbol rather than '$<$', which could have been defined by $\xi < \zeta =_{\text{def}} \exists v(Sv + \xi = \zeta)$.

Since it so greatly helps readability, we'll henceforth make very free use of '$\leq$' as an abbreviatory symbol inside formal arithmetics.

5.8 Why Robinson Arithmetic is interesting

Given it can't even prove $\forall x(0 + x = x)$, Q is evidently a *very* weak theory of arithmetic. Even so, despite its great shortcomings, Q does have some very nice properties.

---

[6] Fine print: choose the quantified variable to avoid any clash of variables with the substituted terms.

As we saw, it can capture the particular decidable relation that obtains when one number is at least as big as another. And in fact, we can now announce a rather surprising general result:

Theorem 17. Q *can capture* all *decidable numerical properties – i.e. it is sufficiently strong in the sense of Defn 19.*

That might initially seem *very* unexpected, given Q's weakness. But remember, 'sufficient strength' was defined as a matter of being able to *case-by-case* prove enough wffs about decidable properties of individual numbers. It turns out that Q's hopeless weakness at proving *generalizations* doesn't stop it from proving enough facts about *particular* numbers.

So that's why Q is especially interesting – it is one of the very weakest arithmetics which is sufficiently strong, and it was isolated by Raphael Robinson 1952 for just that reason.[7] So Q is one of the weakest arithmetics for which Gödelian proofs of incompleteness can be run. Suppose, then, that a theory is formally axiomatized, consistent and can prove everything Q can prove (those do indeed seem very modest requirements). Then what we've just announced and promised can be proved is that any such theory will be sufficiently strong. And therefore e.g. Theorem 8 will apply – any such theory will be incomplete.

However, we can only prove the announced Theorem 17 that Q *does* have sufficient strength if and when we have a quite general theory of effective decidability to hand. And as we said at the end of the previous chapter, we don't want to get embroiled yet in developing that theory. So what we *will* be proving quite soon (in Chapter 9) is a somewhat weaker claim about Q. We'll show that it can capture all so-called 'primitive recursive' properties, where these form a large and very important subclass of the decidable properties. This major theorem will be a crucial load-bearing part of our proofs of various Gödel style incompleteness theorems: it means that Q gives us 'the modest amount of arithmetic' needed for a version of Theorem 2.

But before we get round to showing all that, we are first going to take a look at a *much* richer arithmetic than Q, namely PA.

---

[7]We can't say that Q is *the* weakest sufficiently strong arithmetic. Robinson also isolated another very weak but sufficiently strong arithmetic which neither contains nor is contained in Q. But Robinson's other theory isn't finitely axiomatized, so it is usual to focus on the prettier and finitely axiomatized Q.

6 First-order Peano Arithmetic

The previous chapter introduced two weak theories of arithmetic, BA and Q. In this chapter – jumping over a whole family of intermediate-strength theories – we introduce a *much* richer first-order theory of arithmetic, PA. It's what you get by adding a generous *induction principle* to Q. But what's that?

6.1 Mathematical induction: the very idea

Here is the basic idea we need:

> Whatever numerical property we take, if we can show that (i) zero has that property, and also show that (ii) this property is always passed down from any number n which has it to its successor Sn, then this is enough to show (iii) the property is had by *all* numbers.

This is the key informal *principle of induction*, and is a standard method of proof for establishing arithmetical generalizations.[1]

It is plainly a sound rule, guaranteed by the structure of the natural number series. Why? If a property is possessed by zero and then is passed down from each number that has it to its successor, then it must be possessed by one; and then by two; and then by three; and so on. It will therefore percolate down to any given number – since you can get to any number by starting from zero and repeatedly adding one (there are no stray numbers, outside that sequence of zero and its successors).

6.2 The induction axiom, the induction rule, the induction schema

The intuitive idea, then, is that for any property of numbers, if zero has it and it is passed from one number to the next, then all numbers have it. How are we going to implement this idea in a formal theory of arithmetic?

We've just expressed the intuitive idea as a generalization over properties of numbers. Hence to frame a corresponding formal version, it might seem natural to use a formalized language that enables us to generalize not just over numbers

---

[1] If you are at all hazy about this, you can start by looking at the Wikipedia article on 'Mathematical induction', or (better!) look at the chapter with the same title in Daniel J. Velleman's *How to Prove It* (CUP).

but over properties of numbers. This means it might seem natural to use a language with *second-order* quantifiers. That is to say, we not only have first-order quantifiers running over all the numbers, but also a further sort of quantifier which runs over all properties-of-numbers. In such a language, we could state a second-order

Defn. 22. Induction Axiom.

$$\forall X(\{X0 \wedge \forall x(Xx \rightarrow XSx)\} \rightarrow \forall xXx)$$

(Predicates are conventionally written upper case: so too for variables that are to occupy predicate position.) You can read this Axiom as saying " for any property X, given that 0 has X, and given that if a number has X so does its succcessor, then *every* number has property X."

Seemingly natural though this might be, however, we will be focusing on formal theories whose logical apparatus involves only regular first-order quantification. This isn't due to some perverse desire to work with one hand tied behind our backs. It is because there are troublesome questions about using second-order logic. For a start, there are technical issues: second-order consequence (at least on the natural understanding) can't be captured in a nice formalizable logical system: so theories using a full second-order logic aren't effectively axiomatizable. And then there are more philosophical issues: just how well do we really understand the idea of quantifying over 'all properties of numbers'? Is that a determinate totality which we can quantify over? We don't want to tangle with these worries here and now.

However, if we don't have second-order quantifiers available to range over properties of numbers, how can we handle induction in a first-order setting? Well, one way is to adopt the following rule of inference:

Defn. 23. Induction Rule. *For any suitable open wff $\varphi(x)$ of our arithmetical language, given $\varphi(0)$ and $\forall x(\varphi(x) \rightarrow \varphi(Sx))$, we can infer $\forall x\varphi(x)$.*

We will discuss what counts as 'suitable' in the next section – and then return to the issue again in §7.5.

Alternatively, we can trade in this induction rule for a general axiom schema, and say:

Defn. 24. Induction Schema. *For any suitable open wff $\varphi(x)$ of our arithmetical language, the corresponding instance of this schema*

$$\{\varphi(0) \wedge \forall x(\varphi(x) \rightarrow \varphi(Sx))\} \rightarrow \forall x\varphi(x)$$

can be taken as an axiom.

Given $\varphi(0)$ and $\forall x(\varphi(x) \rightarrow \varphi(Sx))$, we can either apply the Rule to deduce $\forall x\varphi(x)$, or we can take the corresponding instance of the Schema and then apply modus ponens to deduce the same conclusion (assuming our theory can handle conjunctions and conditionals!). The effect will be just the same either way. We'll take the schematic route.

41

6.3 Being generous with induction?

Suppose then that we start again from the first-order arithmetic Q, and aim to build a richer theory in its language L_A by adding induction, e.g. by adopting all the axioms which are 'suitable' instances of the Induction Schema. But what makes for a 'suitable' predicate for use in an instance of the Schema?

Consider *any* open wff $\varphi(x)$ of L_A. This will be built from no more than the constant term '0', the familiar successor, addition and multiplication functions, plus identity and other logical apparatus. Therefore – you might very well suppose – it ought to express a perfectly determinate arithmetical property (even if, in the general case, we can't always decide whether a given number n has the property or not). *So why not be generous and allow any open L_A wff at all to be suitable for substitution in the Induction Schema?*

Here's a positive argument for generosity. Suppose for a moment $\varphi(x)$ has no free variables other than 'x'. Then a corresponding instance of the Induction Schema will only allow us to derive $\forall x \varphi(x)$ when we can *already* prove the corresponding (i) $\varphi(0)$ and also can prove (ii) $\forall x(\varphi(x) \rightarrow \varphi(Sx))$. But if we can *already* prove (i) and (ii), we can already prove each and every one of $\varphi(0)$, $\varphi(S0)$, $\varphi(SS0)$, ... However, there are no 'stray' numbers which aren't denoted by some numeral; so that means that we can prove of each and every number, taken separately, that φ is true of it. What more can it possibly take for φ to express a genuine property that indeed holds for every number, so that (iii) $\forall x \varphi(x)$ is true?

If that reasoning is correct, it seems that we can't overshoot by allowing as axioms instances of the induction schema for *any* open wff φ of L_A with one free variable. The only *usable* instances from our generous range of induction axioms will be those where we can prove the antecedents (i) and (ii) of the relevant conditionals: and in those cases, we will surely have every reason to accept the consequents (iii) too.

In fact, we will officially extend the 'suitable' candidates for φ a step further. We will allow uses of the inference rule where $\varphi(x)$ also has slots for additional variables dangling free (these variables are then simply carried along for the ride, so to speak). Equivalently, we will take the induction axioms to be instances of the induction schema where the expression substituted for $\varphi(x)$ can have other variables dangling free.[2]

6.4 First-order Peano Arithmetic introduced

Suppose then that we are generous with induction and agree that *any* open wff of L_A is suitable for use in an instance of the induction schema. This means moving on from Q, and jumping over a range of possible intermediate theories, to adopt the much richer theory of arithmetic that we can briskly define as follows:

---

[2] For more explanation, see *IGT2*, §§9.3 and 12.1. But we needn't fuss here about elaborating this point.

Defn. 25. PA – First-order Peano Arithmetic[3] – *is the theory with a standard first-order logic whose language is L_A and whose axioms are those of* Q *plus all instances of the induction schema that can be constructed from open wffs of L_A.*

Like BA then, PA as presented here has an infinite number of axioms. However that's fine: it is plainly still decidable whether any given wff has the right shape to be one of the new axioms, so this is still a legitimate formalized theory.[4]

Let's have three initial examples of what we can formally prove using induction. First, we'll check that we have plugged the particular gap we noted in Q. Recall: Q has $\forall x(x + 0 = x)$ as an axiom, so that's trivially a theorem of the theory; but it feebly can't prove $\forall x(0 + x = x)$. But PA can. We just put $0 + x = x$ for $\varphi(x)$, prove $\varphi(0)$ (trivial!), prove $\forall x(\varphi(x) \to \varphi(Sx))$, and use induction to conclude $\forall x \varphi(x)$. Spelling that out in plodding detail:

1.	$0 + 0 = 0$	Instance of Q's Axiom 4
2.	$0 + a = a$	Supposition
3.	$S(0 + a) = Sa$	From 2 by the identity laws
4.	$0 + Sa = S(0 + a)$	Instance of Q's Axiom 5
5.	$0 + Sa = Sa$	From 3, 4
7.	$0 + a = a \to 0 + Sa = Sa$	From 1, 6 by Conditional Proof
8.	$\forall x(0 + x = x \to 0 + Sx = Sx)$	From 7, since a was arbitrary.
9.	$0 + 0 = 0 \land \forall x(0 + x = x \to 0 + Sx = Sx)$	
	From 1, 8	
10.	$\{0 + 0 = 0 \land \forall x(0 + x = x \to 0 + Sx = Sx)\} \to \forall x(0 + x = x)$	
	Instance of Induction Schema	
11.	$\forall x(0 + x = x)$	From 9, 10 by Modus Ponens

For a second example, let's show that PA proves $\forall x(x \neq Sx)$. Just take $\varphi(x)$ to be $x \neq Sx$. Then PA trivially proves $\varphi(0)$ because that's Q's Axiom 1. And PA also proves $\forall x(\varphi(x) \to \varphi(Sx))$ by contraposing Axiom 2. And then an induction axiom tells us that if we have both $\varphi(0)$ and $\forall x(\varphi(x) \to \varphi(Sx))$ we can deduce $\forall x \varphi(x)$, i.e. no number is a self-successor. It's as simple as that.

Yet this trivial little result is worth noting when we recall our deviant interpretation which makes the axioms of Q true while making $\forall x(0 + x = x)$ false: that interpretation featured Kurt Gödel himself added to the domain as a rogue self-successor. A bit of induction, however, rules out self-successors.

A third observation. PA allows, in particular, induction for the formula

$$\varphi(x) : \quad (x \neq 0 \to \exists y(x = Sy)).$$

---

[3]The name is conventional. Giuseppe Peano did publish a list of axioms for arithmetic in 1889. But they weren't first-order, only explicitly governed the successor relation, and – as Peano acknowledged – had already been stated by Richard Dedekind.

[4]PA as we've presented it has an infinite number of axioms: but can we find a finite bunch of axioms for the theory, i.e. a finite set of axioms with the same consequences? No. First-order Peano Arithmetic is not finitely axiomatizable. That's *not* an easy result though!

43

But now note that the corresponding $\varphi(0)$ is a trivial theorem. $\forall x \varphi(Sx)$ is an equally trivial theorem, and that logically entails $\forall x(\varphi(x) \to \varphi(Sx))$. So we can use an instance of the Induction Schema inside PA to derive $\forall x \varphi(x)$. But that's just Axiom 3 of Q. So our initial presentation of PA – as explicitly having all the Axioms of Q – involves a certain redundancy.

6.5 Summary overview of PA

Given its very natural motivation, PA is the benchmark axiomatized first-order theory of basic arithmetic. Just for neatness, then, let's bring together all the elements of its specification in one place. First, to repeat, the *language* of PA is L_A, a first-order language whose non-logical vocabulary comprises just the constant '0', the one-place function symbol 'S', and the two-place function symbols '+', '×'. The built-in interpretation for L_A gives those symbols their standard interpretation and takes the quantifiers to run over the natural numbers.

Second, PA's deductive *proof system* is some standard version of classical first-order logic with identity. The differences between various presentations of first-order logic of course don't make a difference to what sentences can be proved in PA. We just sketched a proof using a Fitch-style system. But it is convenient, however, to fix officially on a Hilbert-style axiomatic system for later metalogical work theorizing about the theory.

And third, its non-logical *axioms* – eliminating the redundancy we just noted from our original specification – are the following sentences:

Axiom 1. $\quad \forall x(0 \neq Sx)$

Axiom 2. $\quad \forall x \forall y(Sx = Sy \to x = y)$

Axiom 3. $\quad \forall x(x + 0 = x)$

Axiom 4. $\quad \forall x \forall y(x + Sy = S(x + y))$

Axiom 5. $\quad \forall x(x \times 0 = 0)$

Axiom 6. $\quad \forall x \forall y(x \times Sy = (x \times y) + x)$

plus every instance (or, if you prefer axioms to be sentences without free variables, the closure of every instance) of the following

Induction Schema $\quad [\varphi(0) \wedge \forall x(\varphi(x) \to \varphi(Sx))] \to \forall x \varphi(x)$

where $\varphi(x)$ is an open wff of L_A that has 'x' free.

6.6 What can PA prove?

Even BA is good at proving quantifier-free equations. Q adds some ability to prove quantified wffs. We have so far noted just three additional quantified theorems that PA can prove. Though it gets tedious, more exploration will reveal that other familiar and not-so-familiar basic truths about the successor, addition, multiplication functions and about the ordering relation (as defined in §5.7) are provable in PA using induction. So how much more can PA prove?

A great deal! In fact, so much that we might reasonably have hoped – at least before we'd heard of Gödel's incompleteness theorems – that PA would turn out to be a *complete* theory that indeed pins down all the truths of L_A.

Here is something else that would have encouraged this false hope, pre-Gödel. Suppose we define the language L_P to be L_A without the multiplication sign. Take P – so-called Presburger Arithmetic – to be the theory couched in the language L_P, whose axioms are Q's now familiar axioms for successor and addition, plus (the universal closures of) all instances of the induction schema that can be formed in the language L_P. In short, P is PA minus multiplication. *Then* P *is a negation-complete theory of successor and addition*. (We are not going to be able to prove that here – the argument uses a standard model-theoretic method called 'elimination of quantifiers' which isn't hard, and was known in the 1920s, but it would just take too long to explain.)

So the situation is as follows, and was known before Gödel proved his incompleteness theorem. (i) There is a complete formal axiomatized theory BA whose theorems are all the truths about successor, addition and multiplication expressible in the quantifier-free language L_B. (ii) There is another complete formal axiomatized theory – equivalent to PA minus multiplication – whose theorems are exactly the first-order truths expressible using just successor and addition. Against this background, Gödel's result that adding multiplication in order to get full PA gives us a theory which is incomplete and incompletable (if consistent) comes as a rather nasty surprise. It wasn't obviously predictable that adding multiplication would make all the difference. Yet it does. As we've said before, as soon we have an arithmetic as strong as Q which has multiplication as well as addition, we get incompleteness.

And by the way, it isn't that a theory of multiplication must in itself be incomplete. In 1929, assuming we have a complete logic, Thoralf Skolem showed that there is a complete theory for the truths expressible in a suitable first-order language with multiplication but lacking addition or the successor function. So, when quantifiers are in play, why does putting multiplication together with addition and successor produce incompleteness? The answer will emerge shortly enough, but pivots on the fact that even a weak first-order arithmetic like Q with all three functions available can express/capture *all* 'primitive recursive' functions. But we'll have to wait until the next-but-one chapter to explain what *that* means.

6.7 Non-standard models of PA?

We saw that Q has 'a non-standard model', i.e. there is a deviant unintended interpretation that still makes the axioms of Q true. Let's finish the chapter by asking whether PA similarly has non-standard models; does it have deviant unintended interpretations that still make *its* axioms all true?

Yes. Assuming PA is true of the natural numbers (and so is consistent), the Löwenheim-Skolem theorem tells us that it must have non-standard models of all infinite sizes. So PA doesn't pin down uniquely the structure of the natural numbers. Indeed, even if we assume that we are looking for a *countable* model – i.e. a model whose elements could in principle be numbered off – there can be non-standard countable models of PA. A standard compactness argument shows this.[5]

I'll finish this chapter by giving a speedy proof of that last claim. But this is an optional extra, assuming you know just a bit about compactness arguments. But if you don't, then no matter – nothing in later chapters depends on you knowing this proof.

Suppose, then, that we add to the language of PA a new constant c, and add to the axioms of PA the additional axioms $c \neq 0$, $c \neq \bar{1}$, $c \neq \bar{2}$, ..., $c \neq \bar{n}$, ... Evidently each *finite* subset of the axioms of the new theory has a model (assuming PA is consistent and has a model). Just take the intended standard model of arithmetic and interpret c to be some number greater than the maximum n for which $c \neq \bar{n}$ is in the given finite suite of axioms.

Since each finite subset of the infinite set of axioms of the new theory has a model, the compactness theorem tells us the whole theory must have a model. And then, by the downward Löwenheim-Skolem theorem there will be in particular a *countable* model of this theory, which contains a zero, its successors-according-to-the-theory, and rogue elements including the denotation of c. Since this structure is a countable model of PA-plus-some-extra-axioms it is, a fortiori, a countable model of PA, and must be distinct from the standard one as its domain has more than just the zero and its successors.[6]

---

[5] The L-S theorem is a standard result at the very beginnings of model theory, explained in any mathematical logic text. Likewise for the compactness theorem. We can't pause over them here.

[6] "OK: that was smart! But can you now describe one of these countable-but-weird models of PA? In particular, what do the interpretations of the successor, addition and multiplication functions now look like?" The relevant functions take some effort to describe. For *Tennenbaum's Theorem* tells us that, for any non-standard model of PA, the interpretations of the addition and the multiplication functions can't be nice computable functions, and so can't be arithmetical functions that we can give a familiar sort of description of. But pursuing this further would take us too far off-piste.

We might note, though, that our compactness argument shows that there are non-standard countable models without assuming that PA is negation incomplete. But if we already have the incompleteness theorem to hand, we can get the same conclusion another way, again appealing to a smidgin of model theory. Suppose Theorem 1 applies, so that there will be a true sentence G_{PA} of basic arithmetic such that $PA \nvdash G_{PA}$ and $PA \nvdash \neg G_{PA}$. That means both $PA + \neg G_{PA}$ and $PA + G_{PA}$ are consistent. So both $PA + \neg G_{PA}$ and $PA + G_{PA}$ will have countable models (by completeness and the Löwenheim-Skolem theorem); these two countable models will both be models of PA, but can't be isomorphic because they give different verdicts on G_{PA}. Hence there are countable models of PA which aren't isomorphic, and so they can't both be (isomorphic to) the standard model.

7 Quantifier complexity

Wffs of the language L_A come in different degrees of *quantifier complexity*. We can distinguish, for a start, so-called Δ_0, Σ_1, and Π_1 wffs. Later, in §11.1, we will note that the standard Gödel sentence that sort-of-says 'I am unprovable' is a Π_1 wff. This is important – it means that, while the Gödel sentence might be very long and messy, there is also a good sense in which it is logically really quite simple. Why? What is a Π_1 wff? This short chapter explains.

7.1 Q knows about bounded quantifications

We often want to say that all/some numbers less than or equal to some bound have a particular property. We can express such claims in formal arithmetics like Q and PA by using wffs of the shape $\forall x(x \leq \tau \to \varphi(x))$ and $\exists x(x \leq \tau \wedge \varphi(x))$, where $x \leq \tau$ is just short for $\exists v(v + x = \tau)$ (see §5.7), and τ can stand in for any term (not just some numeral) so long as it doesn't contain v free. It is standard to further abbreviate such wffs by $(\forall x \leq \tau)\varphi(x)$ and $(\exists x \leq \tau)\varphi(x)$ respectively.

Now note that we have easy results like these:

1. For any n, $Q \vdash \forall x(\{x = \overline{0} \vee x = \overline{1} \vee \ldots \vee x = \overline{n}\} \leftrightarrow x \leq \overline{n})$.

2. For any n, if $Q \vdash \varphi(\overline{0}) \wedge \varphi(\overline{1}) \wedge \ldots \wedge \varphi(\overline{n})$, then $Q \vdash (\forall x \leq \overline{n})\varphi(x)$.

3. For any n, if $Q \vdash \varphi(\overline{0}) \vee \varphi(\overline{1}) \vee \ldots \vee \varphi(\overline{n})$, then $Q \vdash (\exists x \leq \overline{n})\varphi(x)$.

Such results show that Q – and hence a stronger theory like PA – 'knows' that bounded universal quantifications (with fixed number bounds) behave like finite conjunctions, and that bounded existential quantifications (with fixed number bounds) behave like finite disjunctions. Later we will allow ourselves to appeal to simple results like these, without proof.

7.2 Δ_0 wffs

Let's say that

Defn. 26. *An L_A wff is Δ_0 iff it can be built up from the non-logical vocabulary of L_A plus $\leq$ (defined as before), using the familiar propositional connectives, the identity sign, but only bounded quantifications.*

So, a Δ_0 wff is just like a quantifier-free L_A wff, except that we are now allowed the existential quantifiers used in defining occurrences of $\leq$, and we can allow ourselves to wrap up some finite conjunctions into bounded universal quantifications, and similarly wrap up some finite disjunctions into bounded existential quantifications.

It should be no surprise to hear this:

Theorem 18. *We can effectively decide the truth-value of any Δ_0 sentence.*

We won't give a full-dress proof. But, roughly speaking, we can unpack bounded quantifications into conjunctions or disjunctions (perhaps in a number of stages, if bounded quantifiers are nested one inside another). And then we are left with an equivalent wff built up using propositional connectives from basic expressions of the form $\sigma = \tau$ and $\sigma \leq \tau$ (for quantifier-free σ and τ) – and we can compute the truth values of such basic expressions.

Since we can mechanically decide whether $\varphi(\bar{n})$ when that is Δ_0, this means that we can mechanically determine whether a Δ_0 open wff $\varphi(x)$ is satisfied by a given number n. In other words, a Δ_0 open wff $\varphi(x)$ will express a decidable property of numbers. Likewise a Δ_0 open wff $\varphi(x, y)$ will express a decidable numerical relation.

Now, since (i) Theorem 13 tells us that even Q can correctly decide all quantifier-free L_A sentences, (ii) Theorem 16 tells us that Q also knows about the relation $\leq$, and (iii) Q knows that quantifications with fixed number bounds behave just like conjunctions/disjunctions, the next result won't be a surprise either:

Theorem 19. Q *(and hence* PA*) can correctly decide all Δ_0 sentences.*

Again, we won't spell out the argument here.[1]

7.3 Σ_1 and Π_1 wffs

We next say that

Defn. 27. *An L_A wff is Σ_1 if it is (or is logically equivalent to) a Δ_0 wff preceded by zero, one, or more unbounded existential quantifiers. And a wff is Π_1 if it is (or is logically equivalent to) a Δ_0 wff preceded by zero, one, or more unbounded universal quantifiers.*

As a mnemonic, it is worth remarking that 'Σ' in the standard label 'Σ_1' comes from an old alternative symbol for the existential quantifier, as in $\Sigma x F x$ – that's a Greek 'S' for '(logical) sum'. Likewise the 'Π' in 'Π_1' comes from the corresponding symbol for the universal quantifier, as in $\Pi x F x$ – that's a Greek 'P' for '(logical) product'. And the subscript '1' in 'Σ_1' and 'Π_1' indicates that we are

---

[1]Enthusiasts can see the proof of Theorem 11.2 in *IGT2*.

dealing with wffs which start with *one* block of similar quantifiers, respectively existential quantifiers and universal quantifiers.[2]

So a Σ_1 wff says that some number (pair of numbers, etc.) satisfies the decidable condition expressed by its Δ_0 core; likewise a Π_1 wff says that every number (pair of numbers, etc.) satisfies the decidable condition expressed by its Δ_0 core.

To check understanding, pause to make sure you understand why

1. The negation of a Δ_0 wff is still Δ_0.

2. A Δ_0 wff is also Σ_1 and Π_1.

3. The existential quantification of a Σ_1 wff is Σ_1; the universal quantification of a Π_1 wff is Π_1.

4. The negation of a Σ_1 wff is Π_1; the negation of a Π_1 wff is Σ_1.

(For (4), recall the rules for exchanging the order of quantifiers and negations!) And let's note the following easy result:

Theorem 20. Q *can prove any true Σ_1 sentence (is 'Σ_1-complete').*

Proof. Take, for example, a sentence of the type $\exists x \exists y \varphi(x, y)$, where $\varphi(x, y)$ is Δ_0. If this sentence is true, then for some pair of numbers m, n, the Δ_0 sentence $\varphi(\overline{m}, \overline{n})$ must be true. But then by Theorem 19, Q proves $\varphi(\overline{m}, \overline{n})$ and hence $\exists x \exists y \varphi(x, y)$, by existential introduction.

Evidently the argument generalizes for any number of initial quantifiers, which shows that Q proves all truths which are (or are provably-in-Q equivalent to) some Δ_0 wff preceded by one or more unbounded existential quantifiers.   ⊠

7.4   A remarkable corollary

Our last theorem looks entirely straightforward and unexciting, but it has an immediate corollary which is much more interesting:

Theorem 21. *If T is a consistent theory which includes* Q, *then every Π_1 sentence that it proves is true.*

Proof. Suppose T proves a *false* Π_1 sentence φ. Then $\neg\varphi$ will be a *true* Σ_1 sentence. But in that case, since T includes Q and so is 'Σ_1-complete', T will also prove $\neg\varphi$, making T inconsistent. Contraposing, if T is consistent, any Π_1 sentence it proves is true.   ⊠

---

[2] Just for the record, we can keep on going, to consider wffs with greater and greater quantifier complexity. So, we say a Π_2 wff is (or is logically equivalent to) one that starts with *two* blocks of quantifiers, a block of universal quantifiers followed by a block of existential quantifiers followed by a bounded kernel. Likewise, a Σ_2 wff is (equivalent to) one that starts with two blocks of quantifiers, a block of existential quantifiers followed by a block of universal quantifiers followed by a bounded kernel. And so it goes, up the so-called *arithmetical hierarchy* of increasing quantifier complexity. But for our purposes, we won't need to consider levels higher up the arithmetical hierarchy.

Which is, in its way, a quite remarkable observation. It means that we don't have to fully *believe* a theory T – i.e. we don't have to accept that all its theorems are *true* on the interpretation built into T's language – in order to use it to establish that some Π_1 arithmetic generalization is true.

For example, with some minor trickery, we can state Fermat's Last Theorem as a Π_1 sentence. And famously, Andrew Wiles has shown how to derive this Π_1 sentence from some *extremely* heavy-duty infinitary mathematics. Now we see, intriguingly, that that this his background mathematical theory does not need to be *true* – whatever exactly that means when things get so very wildly infinitary! It is enough for Wiles's proof successfully to establish that Fermat's Last Theorem is true that his background theory is *consistent*. Remarkable!

7.5 Intermediate arithmetics

We said at the beginning of the previous chapter that, in moving on from the very weak arithmetics BA and Q to consider first-order PA, we were jumping over a whole family of theories of intermediate strength. We can now briefly describe those intermediate theories: they are the ones we get by restricting the quantifier complexity of suitable instances of the induction schema.

For example, $I\Sigma_1$ is the theory we get by taking the first six axioms of PA (§6.5) plus every instance of the Induction Schema

$$[\varphi(0) \,\wedge\, \forall x(\varphi(x) \rightarrow \varphi(Sx))] \;\rightarrow\; \forall x\varphi(x),$$

where $\varphi(x)$ is now an open Σ_1 wff of L_A.

There is *technical* interest in knowing how much a theory like $I\Sigma_1$ can prove (as we will see in §18.1). But do such theories have any *conceptual* interest? After all, we gave reasons in §6.3 for being generous with induction: we asked, if $\varphi(x)$ expresses a genuine arithmetical property, how can induction fail for $\varphi(x)$?

But which L_A open wffs $\varphi(x)$ (with one free variable) *do* express genuine properties? Previously we took it that they all do (even if, in the general case, we may not be able to decide whether a given number n has the property or not): that is why we said that any such wff $\varphi(x)$ is 'suitable' for appearing in an instance of the Induction Schema (see §6.3 again). But backtrack a moment from that cheerful assumption: suppose you are a *very* stern constructivist who thinks that an expression $\varphi(x)$ only *really* makes sense if it is Δ_0 and so we can effectively decide whether or not it holds true of a given number (or if it is Σ_1 and we can prove it true of a given number when it is). *Then* you can reasonably want to restrict the induction principle to suitable instances using only Δ_0 (or Σ_1) expressions. But it would take us far too far afield even to begin to explore the merits of such proposals here.

Interlude

Let's pause to draw breath and take stock.

1. In Chapter 1 we met the First Incompleteness Theorem in this rough form: a nice enough theory T (which contains the language of basic arithmetic) will always be negation incomplete – there will always be sentences of basic arithmetic it can neither prove nor disprove.

2. We then noted in Chapter 2 that we can cash out the idea of being a 'nice enough' theory in two ways. We can assume T to be *sound*. Or, retreating from that semantic assumption, we can require T to be a *consistent theory which proves a modest amount of arithmetic*. Gödel himself highlights the second version.

3. Of course, we didn't *prove* the Theorem in either version, there at the very outset. However, in Chapter 3, we waved an arm rather airily at the basic strategy that Gödel uses to establish the Theorem – namely we 'arithmetize syntax' (i.e. numerically code up facts about provability in ways that we can express in formal arithmetic) and then construct a Gödel sentence that is (provably) true if and only if is isn't provable.

4. In Chapter 4 we did a bit better, in the sense that we actually gave a *proof* that a consistent, effectively axiomatized, sufficiently strong, formal theory cannot be negation complete.

   The argument was revealing, as it shows that we can get incompleteness results without calling on the arithmetization of syntax and the construction of Gödel sentences. However, the argument depends on the notion of 'sufficient strength' which is defined in terms of the informal notion of a 'decidable property' (a theory, remember, is sufficiently strong if it captures every decidable property of the natural numbers). And the discussion in Chapter 4 doesn't explain how we can sharpen up that informal notion of a decidable property, nor does it explain what a sufficiently strong theory might look like.

5. We need to get less abstract, and start thinking about specific theories of arithmetic. In Chapter 5, as a warm-up exercise, we first looked at BA, the quantifier-free arithmetic of the addition and multiplication of particular numbers. This is a negation-complete and decidable theory – but of

course the theory is only complete, i.e. is only able to decide every sentence constructible in its language, because its quantifier-free language is so very limited. However, if we augment the language of BA by allowing ourselves the usual apparatus of first-order quantification, and replace the schematically presented axioms of BA with their obvious universally quantified correlates (and add in the axiom that every number bar zero is a successor) we get the much more interesting Robinson Arithmetic Q.

Since we are considerably enriching what can be expressed in our arithmetic language while not greatly increasing the power of our axioms, it is no surprise that Q is negation incomplete. And we can prove this without any fancy Gödelian considerations. We can easily show, for example, that Q can't prove either $\forall x(0 + x = x)$ or its negation. Q, then, is a very weak arithmetic. Still, it will turn out to be the 'modest amount of arithmetic' needed to get a syntactic version of the First Theorem to fly. We announced (but of course haven't proved) that even Q is sufficiently strong: which explains why Q turns out to be so interesting despite its weakness.

6. In Chapter 6, we then moved on to introduce first-order Peano Arithmetic PA, which adds to Q a whole suite of induction axioms (every instance of the Induction Schema). Exploration reveals that this theory, in contrast to Q, is very rich and powerful. We might, pre-Gödel, have very reasonably supposed that it is a negation-complete theory of the arithmetic of addition and multiplication. But the theory is still effectively axiomatized, and the First Theorem is going to apply (assuming PA is sound, or is at least consistent). So PA too will turn out to be negation incomplete.

7. There are theories intermediate in strength between Q and PA, theories which have induction axioms but only for wffs up to some degree of quantificational complexity. We will be interested in one such intermediate theory later in these notes (Chapter 18). But the task of Chapter 7 was just to explain this notion of quantificational complexity, and in particular explain what Σ_1 and Π_1 wffs are.

Which brings us up to the current point in these notes. To give a sense of direction, let's next outline where we are going in the next five chapters. (Skip if you don't want spoilers!)

8. The formal theories of arithmetic that we've looked at so far have (at most) the successor function, addition and multiplication built in. But why stop there? Even high-school arithmetic acknowledges many more numerical functions, like the factorial and the exponential.

   Chapter 8 describes a very wide class of such numerical functions, the so-called primitive recursive (p.r.) ones. They are a major subclass of the effectively computable functions.

   We also define the primitive recursive properties and relations – a numerical property/relation is p.r. when some p.r. function can effectively decide when it holds.

9. Chapter 9 then shows that L_A, the language of basic arithmetic, can *express* all p.r. functions and relations. Moreover Q and hence PA can *capture* all those functions and relations too (i.e. case-by-case prove wffs that assign the right values to the functions for particular numerical arguments). So Q and PA, despite having only successor, addition and multiplication 'built in', can actually deal with a vast range of functions (at least in so far as they can 'calculate' the value of the functions for arbitrary numerical inputs).

   Note the link with our earlier talk about 'sufficiently strong theories' (Defn. 19). Those, recall, are theories that can capture all effectively decidable properties of numbers. Well, now we are going to show that PA (indeed, even Q) can capture at least all those effectively decidable properties of numbers which are primitive recursive (a very important class). And we'll find that that's enough for the core Gödelian argument to go through.

10. In Chapter 10 we then introduce again the key idea of the 'arithmetization of syntax' by Gödel-numbering which we first met in §§3.3 and 3.4. Focus on PA for the moment, and fix on a suitable Gödel-numbering. Then we can define various numerical properties/relations such as:

    Wff(n) iff n is the code number of a PA-wff;
    Sent(n) iff n is the code number of a PA-sentence;
    Prf(m, n) iff m is the code number of a PA-proof of the sentence with code number n.

    Moreover – the crucial result – these properties/relations are primitive recursive. Similar results obtain for any sensibly axiomatized formal theory.

11. Since *Prf* is p.r., and the theory PA can express all p.r. relations, we can express some facts about PA-proofs in PA itself. In chapter 11 we use this fact in constructing a Gödel sentence which is true if and only if it is not provable in PA. We can thereby prove the semantic version of Gödel first incompleteness theorem for PA in something close to Gödel's way, assuming PA is sound. The result generalizes to other sensibly axiomatized sound arithmetics that include Q.

12. Then Chapter 12, at last, proves a crucial syntactic version of the First Incompleteness Theorem, again in something close to Gödel's way.

Now read on ...

8 Primitive recursive functions

As we have just announced in the Interlude, the primitive recursive functions form a major subclass of the effectively computable functions (in fact, any computable numerical function you already know about is almost certainly primitive recursive). This chapter explains what these functions are, and proves some elementary results about them.

8.1 Introducing the primitive recursive functions

Let's start by revisiting the basic axioms for *addition* and *multiplication* which we adopted formally in Q and PA. Here again is what the axioms say, but now presented in the style of everyday informal mathematics – and note, everything in this chapter belongs to informal mathematics. So, leaving quantifiers to be understood in the familiar way, and taking the variables to be running over the natural numbers, the principles are:

$$x + 0 = x$$
$$x + Sy = S(x + y)$$

$$x \times 0 = 0$$
$$x \times Sy = (x \times y) + x$$

The first of the pair of equations for addition tells us the result of adding zero to a given number x. The second tells us the result of adding Sy (i.e. adding the successor of y) in terms of the result of adding y. Hence these equations – as we pointed out before – together tell us how to add any of $0, S0, SS0, SSS0, \ldots$, i.e. they tell us how to add *any* number to a given number x. Similarly, the first of the pair of equations for multiplication tells us the result of multiplying by zero. The second equation tells us the result of multiplying by Sy in terms of the result of multiplying by y. Hence these equations together tell us how to multiply a given number x by any of $0, S0, SS0, SSS0, \ldots$, i.e. they tell us how to multiply by *any* number.

Here are two more functions that are familiar from elementary arithmetic. Take the *factorial* function $y!$, where e.g. $4! = 1 \times 2 \times 3 \times 4$. Then the factorial function can be defined by the following two equations:

$$0! = 1$$
$$(Sy)! = y! \times Sy$$

The first equation tells us the conventional value of the factorial function for the argument 0; the second equation tells us how to work out the value of the function for Sy once we know its value for y (assuming we already know about multiplication). So by applying and reapplying the second equation, we can successively calculate $1!, 2!, 3!, 4! \ldots$, as follows:

$$1! = 0! \times 1 = 1$$
$$2! = 1! \times 2 = 2$$
$$3! = 2! \times 3 = 6$$
$$4! = 3! \times 4 = 24$$

And so on and on it goes. Our two-equation definition is properly called a definition because it fixes the value of '$y!$' for all numbers y.

For our next example – this time another two-place function – consider the *exponential*, standardly written in the form 'x^y'. This can be defined by a similar pair of equations:

$$x^0 = S0$$
$$x^{Sy} = (x^y \times x)$$

Again, the first equation gives the function's value for a given value of x when $y = 0$, and – keeping x fixed – the second equation gives the function's value for the argument Sy in terms of its value for y. The equations determine, e.g., that $3^4 = 3 \times 3 \times 3 \times 3 = 81$.

Three comments about our examples so far. (1) Note that in each definition, the second equation fixes the value of a function for argument Sy by invoking the value of the *same* function for argument y. A procedure where we evaluate a function for one input by calling the *same* function for a smaller input is standardly termed 'recursive' – and the particularly simple pattern we've illustrated is called, more precisely, 'primitive recursive'. So our two-clause definitions are examples of *definition by primitive recursion*.[1]

(2) Next note, for example, that $(Sy)!$ is defined as $y! \times Sy$, so it is evaluated by evaluating $y!$ and Sy and then feeding the results of these computations into the multiplication function. This involves, in a word, the *composition* of functions, where evaluating a composite function involves taking the output(s) from one or more functions, and treating these as inputs to another function.

(3) Our four examples can be arranged into two short *chains* of definitions by recursion and functional composition. Working from the bottom up, addition

---

[1] "Surely, defining a function in terms of that very same function is circular!" But of course, that isn't quite what's happening. We are fixing the value of the function for one input in terms of its already-settled value for a *smaller* input: and *that* is not circular. Still, strictly speaking, we can ask for a proof confirming that primitive recursive definitions really do well-define functions: such a proof was first given by Richard Dedekind in 1888.

is defined in terms of the successor function; multiplication is then defined in terms of successor and addition; then the factorial (or, in the second chain, exponentiation) is defined in terms of multiplication and successor.

Here's another short chain of definitions:

$$P(0) = 0$$
$$P(Sx) = x$$

$$x \mathbin{\dot-} 0 = x$$
$$x \mathbin{\dot-} Sy = P(x \mathbin{\dot-} y)$$

$$|x - y| = (x \mathbin{\dot-} y) + (y \mathbin{\dot-} x)$$

'P' signifies the predecessor function (with zero being treated as its own predecessor); '$\mathbin{\dot-}$' signifies 'subtraction with cut-off', i.e. subtraction restricted to the non-negative integers (so $m \mathbin{\dot-} n$ is zero if $m < n$). And $|m - n|$ is the absolute difference between m and n. This time, our third definition doesn't involve recursion, only a simple composition of functions.

These chains of definitions motivate the following initial, quick-and-dirty, way of specifying the p.r. functions:

Defn. 28. *Roughly: a primitive recursive function is one that can be similarly characterized using a chain of definitions by recursion and composition, starting from trivial 'initial functions' like the successor function.*[2]

8.2   Defining the p.r. functions more carefully

On the one hand, I suppose you really ought to read this section! On the other hand, *don't* get lost in the details. All we are trying to do here is to give a more careful presentation of the ideas we have just been sketching, and to elaborate that last rough definition.

We have three things to explain more carefully. (a) We need to tidy up the idea of defining a function by primitive recursion. (b) We need to tidy up the idea of defining a new function by composing old functions. And (c) we need to say more about the 'starter pack' of initial functions which we can use in building up a chain of definitions by primitive recursion and/or composition.

We'll take these steps in turn.

(a)   Consider the recursive definition of the factorial again:

$$0! = 1$$
$$(Sy)! = y! \times Sy$$

This is an example of the following general scheme for defining a one-place function f:

---

[2]The basic idea is there in Dedekind and highlighted by Skolem in 1923. But the modern terminology 'primitive recursion' seems to be due to Rósza Péter in 1934; and 'primitive recursive function' was first used by Stephen Kleene in 1936.

$$f(0) = g$$
$$f(Sy) = h(y, f(y))$$

Here, g is just a number, while h is a two-place function which – crucially – *we are assumed already to know about* prior to the definition of f. Maybe that's because h is an 'initial' function that we are allowed to take for granted; or maybe it's because we've already given recursion clauses to define h; or maybe h is a composite function constructed by plugging one known function into another – as in the case of the factorial, where $h(y, z) = z \times Sy$ (where we take the output from the successor function as one input into the multiplication function).

Likewise, with a bit of massaging, the recursive definitions of addition, multiplication and the exponential can all be treated as examples of the following general scheme for defining two-place functions:

$$f(x, 0) = g(x)$$
$$f(x, Sy) = h(x, y, f(x, y))$$

where now g is a one-place function, and h is a three-place function, again functions that we already know about. Three points about this:

i. To get the definition of addition to fit this pattern, with a unary function on the right of the first equation, we take $g(x)$ to be the trivial *identity function* $I(x) = x$.

ii. To get the definition of multiplication to fit the pattern, $g(x)$ has to be treated as the equally trivial *zero function* $Z(x) = 0$.

iii. Again, to get the definition of addition to fit the pattern, we have to take $h(x, y, z)$ to be the function Sz. As this illustrates, we must allow h not to care what happens to some of its arguments, while operating on some other argument(s). The conventional way of doing this is to help ourselves to some further trivial identity functions that serve to select out particular arguments. For example, the function I_3^3 takes three arguments, and just returns the third of them, so $I_3^3(x, y, z) = z$. Then, in the definition of addition, we can put $h(x, y, z) = SI_3^3(x, y, z)$, so h is defined by composition from initial functions which we can take for granted.

We can now generalize the idea of a definition by recursion from the case of one-place and two-place functions to cover the case of many-place functions. There's a standard notational device that helps to put things snappily: we write $\vec{x}$ as short for the array of k variables $x_1, x_2, \ldots, x_k$ (taking the relevant k to be fixed by context). Then:

Defn. 29. *Suppose that the following holds:*

$$f(\vec{x}, 0) = g(\vec{x})$$
$$f(\vec{x}, Sy) = h(\vec{x}, y, f(\vec{x}, y))$$

Then f is defined from g and h by primitive recursion.

This covers the case of one-place functions $f(y)$ like the factorial if we allow $\vec{x}$ to be empty, in which case $g(\vec{x})$ is a 'zero-place function', i.e. a constant.

(b)   Now to tidy up the idea of definition by composition. The basic idea, to repeat, is that we form a composite function f by treating the output value(s) of one or more given functions $g, g', g'', \ldots$, as the input argument(s) to another function h. For example, we set $f(x) = h(g(x))$. Or, to take a slightly more complex case, we could set $f(x, y, z) = h(g(x, y), g'(y, z))$.

There's a number of equivalent ways of covering the manifold possibilities of compounding multi-place functions. But one natural way is to define what we might call one-at-a-time composition (where we just plug *one* function g into another function h), thus:

Defn. 30. *If $g(\vec{y})$ and $h(\vec{x}, u, \vec{z})$ are functions – with $\vec{x}$ and $\vec{z}$ possibly empty – then f is defined by composition by substituting g into h just if $f(\vec{x}, \vec{y}, \vec{z}) = h(\vec{x}, g(\vec{y}), \vec{z})$.*

We can then think of generalized composition – where we plug more than one function into another function – as just iterated one-at-a-time composition. For example, we can substitute the function $g(x, y)$ into $h(u, v)$ to define the function $h(g(x, y), v)$ by composition. Then we can substitute $g'(y, z)$ into the defined function $h(g(x, y), v)$ to get the composite function $h(g(x, y), g'(y, z))$.

(c)   So far, so good (no one promised the details would be exciting!).

Now, the quick-and-dirty Defn. 28 tells us that the primitive recursive functions are built up by recursion and composition, beginning from some 'starter pack' of trivial basic functions. But which functions are they? We have met all the ones we need:

Defn. 31. *The initial functions are the successor function S, the zero function $Z(x) = 0$ and all the k-place identity functions, $I_i^k(x_1, x_2, \ldots, x_k) = x_i$ for each k, and for each i, $1 \le i \le k$.*

These identity functions are also often called *projection* functions (they 'project' the vector with components $x_1, x_2, \ldots, x_k$ onto the i-th axis).

Let's now put everything together. We informally defined the primitive recursive (as before, 'p.r.' for short[3]) functions as those that can be defined by a chain of definitions by recursion and composition. We can now says more formally:

Defn. 32. *The p.r. functions are the following:*

1. *The successor function S, zero function Z, and all the identity functions I_i^k are p.r.;*

2. *if f can be defined from the p.r. functions g and h by composition, substituting g into h, then f is p.r.;*

---

[3]Terminology alert: some authors writing in this area use 'p.r.' as short for *partial* recursive – a quite different notion!

3. *if f can be defined from the p.r. functions g and h by primitive recursion, then f is p.r.;*

4. *nothing else is a p.r. function.*

(We allow g in clauses (2) and (3) to be zero-place, i.e. be a constant.)

So a p.r. function f is one that *can* be specified by a chain of definitions by recursion and composition, leading back to initial functions. So let's say:

Defn. 33. *A definition chain for the p.r. function f is a sequence of functions $f_0, f_1, f_2, \ldots, f_k$ where each f_j is either an initial function or is defined from previous functions in the sequence by composition or recursion, and $f_k = f$.*

The closure condition (4) in the previous definition means that every p.r. function is required to have a definition chain in this sense (the chain need not be unique, but the function must have at least one to be p.r.) – which sharpens the informal characterization Defn. 28 which we gave at the end of the previous section.

8.3 How to prove a result about all p.r. functions

The point that every p.r. function has a definition chain means that there is a simple method of proving that every p.r. function shares some feature. Suppose that, for some given property P, we can show the following:

P1. The initial functions have property P.

P2. If the functions g and h have property P, and f is defined by composition from g and h, then f also has property P.

P3. If the functions g and h have property P, and f is defined by primitive recursion from g and h, then f also has property P.

Then P1, P2, and P3 together suffice to establish that all primitive recursive functions have property P.

Why? Take any p.r. function f. It must have a definitional chain. Now trek along a definitional chain for f. Each initial function we encounter has property P by P1. By P2 and P3, each definition by recursion or composition which is used in the chain takes us from functions which have property P to another function with property P. So, every function we define as we go along the chain has property P, including the final target function f.

In sum, then: to prove that all p.r. functions have some property P, it suffices to prove the relevant versions of P1, P2 and P3.

For a simple first example, take the property of being a *total function* of the natural numbers, i.e. being a function which outputs a natural number value for any given numerical input. (Example: the function x^2 defined over the natural numbers is total – give it a natural number, and it outputs a natural number; but the function $\sqrt{x}$ defined over the natural numbers is only partial – for input 4 it outputs 2; but 5 has no square root in the natural numbers, and so $\sqrt{5}$ has no value.) Now, the initial functions are, trivially, total functions of numbers, defined for every numerical argument; also, primitive recursion and composition both build total functions out of total functions (why?). Which means that all p.r. functions are total functions, defined for all natural number arguments.

8.4 The p.r. functions are computable

We now show that every p.r. function is effectively computable. That is to say, for any given input(s) we can compute the function's value by a step-by-step algorithmic procedure. Given the general strategy just described, it is enough to show these:

C1. The initial functions are computable.

C2. If f is defined by composition from computable functions g and h, then f is also computable.

C3. If f is defined by primitive recursion from the computable functions g and h, then f is also computable.

For C1 we just remark that the initial functions S, Z, and I_i^k are effectively computable by trivial algorithms. For C2, note that to compute the composition of two computable functions g and h you just feed the output from whatever algorithmic routine evaluates g as input into the routine that evaluates h.

To illustrate C3, return once more to our example of the factorial. Here is its p.r. definition again,

$$0! = 1$$
$$(Sy)! = y! \times Sy$$

The first clause gives the value of the function for the argument 0; then – as we said – you can repeatedly use the second recursion clause to calculate the function's value for $S0$, then for $SS0$, $SSS0$, etc. So the definition encapsulates an algorithm for calculating the function's value for any number, and corresponds exactly to a certain simple kind of computer routine. And obviously the argument generalizes to establish C3.

8.5 Computations for p.r. functions

A high level computer language like Haskell can deal with the two equations giving the p.r. definition of the factorial more or less as they stand. But let's consider a program for computing the factorial which uses a somewhat lower-level language. So compare our recursive definition with the following schematic program which takes a number n as input:

1. $fact := 1$
2. For $y = 0$ to $n - 1$

3. $fact := (fact \times Sy)$
4.   Loop

Here, *fact* is a register that we initially prime with the value of $0!$. Then the program starts looping,[4] updating the contents of *fact* on each iteration. And the crucial thing about executing this kind of *'for' loop* is that the total number of iterations to be run through is given a fixed bound in advance: you count the loops from 0, and after executing a loop, you increment the counter by one on each cycle until you exceed the given bound and exit. So in this case, on loop number y the program replaces the value in the register with Sy times the previous value (we'll assume the computer already knows how to do multiplication). When the program exits the loop after a total of n iterations, the value in the register *fact* will be $n!$.

More generally, for any one-place function f defined by recursion in terms of g and the computable function h, the same program structure always does the trick for calculating $f(n)$. Thus compare the second clause of

$$f(0) = g$$
$$f(Sy) = h(y, f(y))$$

with the corresponding program which takes input n:

1. $func := g$
2.   For $y = 0$ to $n - 1$
3. $func := h(y, func)$
4.   Loop

Given that h is computable, the value of $f(n)$ will be computable using this bounded 'for' loop that terminates with the required value in the register *func*.

Similarly, of course, for many-place functions. For just one example, take the recursive clause of the definition of the addition function: $x + Sy = S(x + y)$. There is a corresponding program which takes as input numbers m and n and terminates with the sum $m + n$ in the register *add*:

1. $add := m$
2.   For $y = 0$ to $n - 1$
3. $add := S(add)$
4.   Loop

In other words, the effect of the definition by recursion can be computed by a bounded 'for' loop.

Now, our mini-program for the factorial calls the multiplication function which can itself be computed by a similar 'for' loop (invoking addition). And addition as we have just seen can be computed by another 'for' loop (invoking the successor). So reflecting the downward chain of recursive definitions

---

[4]Fine print: in the special case when $n = 0$, i.e. when faced with the instruction 'For $y = 0$ to -1, do some stuff', the computer does zero loops – or, if you like, just skips the looping procedure.

61

factorial $\Rightarrow$ multiplication $\Rightarrow$ addition $\Rightarrow$ successor

there's a composite program for the factorial containing *nested* bounded 'for' loops, which ultimately calls the primitive operation of incrementing the contents of a register by one (or other operations like setting a register to zero, corresponding to the zero function, or copying the contents of a register, corresponding to an identity function).

The point obviously generalizes, giving us

Theorem 22. *Primitive recursive functions are effectively computable by a program which invokes a series of (possibly nested) bounded 'for' loops.*

And the crucial thing here, a point we will return to, is that the required looping procedures are each iterated a fixed maximum number of times, set in advance as we first enter the loop (we can allow early exits). Contrast the open-ended searches involved in 'do until' (or 'do while') procedures, where we keep on looping around for as long as it takes, until (or while) some condition is satisfied.

The converse of the last theorem is also true. Suppose we have a program which sets a value for $f(0)$, and then goes into a bounded-in-advance 'for' loop which computes the value of a one-place function $f(n)$ (for $n > 0$), a loop which calls on some already-known function(s) which are used on loop number y (counting from zero) to fix the value of $f(Sy)$ in terms of the value of $f(y)$. This plainly corresponds to a definition by recursion of f. And generalizing,

Theorem 23. *If a function can be computed by a program without open-ended searches – using just bounded 'for' loops for iterative procedures and with the program's 'built in' functions all being p.r. – then the newly defined function will also be primitive recursive.*

In fact we can expand this a bit to allow e.g. conditionally branching computations (as long as the test condition for the branching is itself p.r. computable) – but we needn't fuss about this here.

This gives us a quick way of convincing ourselves that a new function is p.r.: sketch out a routine for computing it and check that the needed looping computations only invoke already known p.r. functions and the number of iterations in any loop is always bounded in advance, so we do not need to set off on any open-ended searches. Then the new function will be primitive recursive.

For a quick example, take the two-place function $gcd(x, y)$ which outputs the greatest common divisor of the two inputs. Evidently, a bounded search through cases is enough to do the trick: at its crudest and most inefficient, we can look in turn at all the numbers up to (and including) the smaller of x and y and see if it divides both. That assures us that $gcd(x, y)$ is p.r. without going through the palaver of actually writing down a suitable definition chain.

8.6   Not all computable numerical functions are p.r.

We have seen that any p.r. function is effectively computable. And most of the ordinary computable numerical functions you already know about from elementary maths are in fact primitive recursive. *But not all effectively computable numerical functions are primitive recursive.*

First, some plausibility considerations. We've just seen that the values of a given primitive recursive function can be computed by a program involving bounded 'for' loops as its main programming structure. Each loop goes through a specified number of iterations, set in advance. However, we *do* allow procedures involving open-ended searches to count as effective computations, even if there is no prior bound on the length of search (so long as we know the search will eventually terminate in a perhaps unknown but finite number of steps). We made essential use of this permission when we showed that negation-complete theories are decidable: we allowed the process *enumerate the theorems and wait to see which of* φ *or* $\neg\varphi$ *turns up* to count as a computational decision procedure.

Standard computer languages of course have programming structures which implement just this kind of unbounded search. Because, in one form or another, as well as bounded 'for' loops, they allow open-ended 'do until' loops (or equivalently, 'do while' loops). In other words, they allow some process to be iterated until a given condition is satisfied – *where no prior limit is put on the number of iterations to be executed.*[5]

If we count what are presented as unbounded searches as computations, then it looks very plausible that not everything computable will be primitive recursive.

True, that is as yet only a plausibility consideration. Our remarks so far leave open the possibility that computations can always somehow be turned into procedures using 'for' loops with a bounded limit on the number of steps. But we can prove that isn't the case:

Theorem 24. *There are effectively computable numerical functions which aren't primitive recursive.*

Proof. The p.r. functions are effectively enumerable. That is to say, there is an effective way of numbering off functions $f_0, f_1, f_2, \ldots$, such that each of the f_i is p.r., and each p.r. function appears somewhere on the list.

Why? A p.r. function is defined by recursion or composition from other functions which are defined by recursion or composition from other functions which are defined ... ultimately in terms of some primitive starter functions. So choose some standard formal specification language for representing these chains of definitions. Then we can effectively generate all possible strings of symbols from this language (by length, then 'in alphabetical order'); and as we go along, we select

---

[5]Programming languages differ as to whether they explicitly mark the difference with distinct instructions, 'for' as against 'do while'/'do until'. But the point of principle remains: there is a difference between cases where the bound to a looping procedure is given in advance, and cases where the procedure is allowed to carry on for an indefinite number of iterations.

the strings that obey the rules for being a definition chain for a p.r. function. This generates a list which effectively enumerates the p.r. functions, repetitions allowed. So now consider the following table:

	0	1	2	3	$\cdots$
f_0	$\underline{f_0(0)}$	$f_0(1)$	$f_0(2)$	$f_0(3)$	$\cdots$
f_1	$f_1(0)$	$\underline{f_1(1)}$	$f_1(2)$	$f_1(3)$	$\cdots$
f_2	$f_2(0)$	$f_2(1)$	$\underline{f_2(2)}$	$f_2(3)$	$\cdots$
f_3	$f_3(0)$	$f_3(1)$	$f_3(2)$	$\underline{f_3(3)}$	$\cdots$
$\cdots$	$\cdots$	$\cdots$	$\cdots$	$\cdots$	$\searrow$

Down the table we list off the p.r. functions f_0, f_1, f_2, ... An individual row then gives the values of a particular f_n for each argument. Let's define the corresponding *diagonal* function, by putting $\delta(n) = f_n(n) + 1$. To compute $\delta(n)$, we just run our effective enumeration of the recipes for p.r. functions until we get to the recipe for f_n. We follow the instructions in that recipe to evaluate that function for the argument n. We then add one. Each step is entirely mechanical. So our diagonal function is effectively computable, using a step-by-step algorithmic procedure.

By construction, however, the function δ can't be primitive recursive. For suppose otherwise. Then δ must appear somewhere in the enumeration of p.r. functions, i.e. be the function f_d for some index number d. But now ask what the value of $\delta(d)$ is. By hypothesis, the function δ is none other than the function f_d, so $\delta(d) = f_d(d)$. But by the initial definition of the diagonal function, $\delta(d) = f_d(d) + 1$. Contradiction.

So we have, as they say, 'diagonalized out' of the class of p.r. functions to define a new function δ which is still effectively computable but not primitive recursive. $\boxtimes$

"But hold on! *Why* is δ not a p.r. function?" Well, consider evaluating $\delta(n)$ for increasing values of n. For each new argument, we will have to evaluate a *different* function f_n for that argument (and then add 1). Evaluating these different functions f_n for input n will call on computations involving loops nested to varying different depths. We have no reason to expect there will be a nice pattern in the successive computations of all the different functions f_n which enables them to be wrapped up into a single p.r. definition. And our diagonal argument in effect shows that this can't be done.

8.7 Defining p.r. properties and relations

We have defined the class of p.r. *functions*. Finally in this chapter, we extend the scope of the idea of primitive recursiveness and introduce the ideas of *p.r.*

decidable (numerical) properties and *relations*.

Now, quite generally, we can tie together talk of functions and talk of properties and relations by using the notion of a *characteristic function*:

Defn. 34. *The characteristic function of the numerical property P is the one-place function c_P such that if m is P, then $c_P(m) = 0$, and if m isn't P, then $c_P(m) = 1$.*

The characteristic function of the two-place numerical relation R is the two-place function c_R such that if m is R to n, then $c_R(m, n) = 0$, and if m isn't R to n, then $c_R(m, n) = 1$.

And similarly for many-place relations. The choice of values for the characteristic function is, of course, entirely arbitrary: any pair of distinct numbers would do. Our choice is supposed to be reminiscent of the familiar use of 0 and 1, one way round or the other, to stand in for *true* and *false*. And our selection of 0 rather than 1 for *true* follows Gödel.

The numerical property P partitions the numbers into two sets, the set of numbers that have the property and the set of numbers that don't. Its corresponding characteristic function c_P also partitions the numbers into two sets, the set of numbers the function maps to the value 0, and the set of numbers the function maps to the value 1. And these are the *same* partition. So in a good sense, P and its characteristic function c_P contain exactly the same information about a partition of the numbers: hence we can move between talk of a property and talk of its characteristic function without loss of information. Similarly, of course, for relations (which partition pairs of numbers, etc.). And we can use this link between properties and relations and their characteristic functions in order to carry over ideas defined for functions and apply them to properties/relations.

For example, without further ado, we now extend the idea of primitive recursiveness to cover properties and relations:

Defn. 35. *A p.r. decidable property is a property with a p.r. characteristic function, and likewise a p.r. decidable relation is a relation with a p.r. characteristic function.*

By way of casual abbreviation, we'll fall into saying that p.r. decidable properties and relations are themselves (simply) p.r.

For a quick example, consider the property of being a *prime* number. Take the characteristic function $pr(n)$ which has the value 0 when n is prime, and 1 otherwise. Now just note that we can evidently compute $pr(n)$ just using 'for' loops (we just do a bounded search through numbers less than n – indeed, no greater than $\sqrt{n}$ – and if we find a divisor of n other than 1, return the value 1, and otherwise return the value 0). So the property of being prime is p.r.

9 Expressing and capturing the primitive recursive functions

Addition can be defined in terms of repeated applications of the successor function. Multiplication can be defined in terms of repeated applications of addition. The exponential and factorial functions can be defined, in different ways, in terms of repeated applications of multiplication. There's already a pattern emerging here! And the main task of the last chapter was to get clear about this pattern.

So first we said more about this way of defining one function in terms of repeated applications of another function. Tidied up, this becomes the idea of *defining a function by primitive recursion* (Defn. 29). Then we explained the idea of giving a definitional chain which defines a function by primitive recursion and/or composition from other functions which we define by primitive recursion and/or composition from other functions, and so on down, until we bottom out with the successor function and other trivia. Tidied up, this gives us the idea of a *primitive recursive function*, i.e. one that can be characterized by such a chain of definitions (Defn. 32).

We noted three key facts:

1. Every p.r. function is effectively computable – moreover it is computable using only bounded 'for' loops, without open-ended searches using 'do until' loops. That's Theorem 22.

2. Conversely, if a numerical function can be computed from simpler p.r. functions without open-ended searches (using only bounded'for' loops for iterative procedures), then it too is primitive recursive. That's Theorem 23.

3. But not every intuitively computable numerical function is primitive recursive. That's Theorem 24.

So the situation is now this. In Chapters 5 and 6, we introduced some *formal* arithmetics with just three functions – successor, addition, multiplication – built in. But we have now reminded ourselves that ordinary *informal* arithmetic talks about many more elementary computable functions like the factorial, the exponential, and so on: and we generalized the sort of way these functions can be defined to specify the whole class of primitive recursive functions. A gulf seems to have opened up, then, between the extreme modesty of the resources of our formal theories (including the strongest so far, PA) and the great richness of the world of p.r. functions (and we know that those aren't even all the computable arithmetical functions).

9.1   Bridging the divide

The aim of this chapter is to show that the theories Q and PA, despite the modesty of their built-in resources, reach a lot further than you might expect. In particular, we state two key results, Theorems 26 and 28. The first will tell us that

> The language L_A can express all primitive recursive functions.

And there's more: looking at the proof of that result we find that in fact L_A can express any p.r. function using a Σ_1 wff – i.e. by using a wff of low quantifier complexity. The second of our theorems then tells us that

> The theory Q – and hence any stronger theory like PA – can capture any p.r. function.

Again, the capturing can be done by using the same Σ_1 wffs.

Now, the *ideas* involved in proving these two major theorems are not particularly difficult. But working through the proofs takes quite a bit of care and patience. It is up to you how much detail you want to take aboard. For enthusiasts, we'll give most of a proof of Theorem 26 (there are some nice ideas involved); but we will only outline what it takes to prove Theorem 28. Don't get bogged down, and do feel free to skim and skip on to the next chapter.

9.2 L_A can express the factorial function

In this section, as a warm-up exercise, we are going to show that L_A – despite having only successor, addition and multiplication built in – can *express* the factorial function. That is to say, we can construct an L_A wff $\mathsf{F}(\mathsf{x}, \mathsf{y})$ such that, for any particular m and n, $\mathsf{F}(\overline{m}, \overline{n})$ if and only if $n = m!$.

Then in the next section we'll use the same key trick in showing that L_A can express *any* p.r. function at all. But it will be much easier to follow the general argument if you first meet the 'β-function trick' when deployed in a simple case in this section. The details are fiddly: so we'll take things in five stages.

(a)   Consider, then, the p.r. definition of the factorial function again:

$$0! = 1$$
$$(Sx)! = x! \times Sx$$

Now, think of this definition in the following way: for any x, it tells us how to construct a sequence of numbers $0!, 1!, 2!, \ldots, x!$, where we move from the i-th member of the sequence (counting from zero) to the next by multiplying by Si. Or putting it a bit more abstractly, suppose that for numbers x and y,

1. There is a sequence of numbers $k_0, k_1, \ldots, k_x$ such that: $k_0 = 1$, and if $i < x$ then $k_{Si} = k_i \times Si$, and $k_x = y$.

Then this is equivalent to saying that $y = x!$.

So the question of how to reflect the p.r. definition of the factorial inside L_A can be parlayed into the following question: how can we express facts about *finite sequences of numbers* using the limited resources of L_A?

(b)   Use numerical codes! Suppose we can wrap up a finite sequence into a single *code number* c, and then have a two-place *decoding function*, call it simply *decode*, such that if you give *decode* the code c and the index i, the function spits out the i-th member of the sequence which c codes. In other words, suppose that, when c is the code number for the sequence $k_0, k_1, \ldots, k_x$, then $decode(c, i) = k_i$.

If we can find such a coding scheme, then we can rewrite (1) as follows, talking about a code number c instead of the sequence $k_0, k_1, \ldots, k_x$, and writing $decode(c, i)$ instead of k_i:

2. There is a code number c such that: $decode(c, 0) = 1$, and if $i < x$ then $decode(c, Si) = decode(c, i) \times Si$, and $decode(c, x) = y$.

This way, if a suitable *decode* function can indeed be expressed in L_A, then we can say $y = x!$ in L_A. Great! So can we do this coding trick?

(c)   To link up with Gödel's own version of the trick, let's liberalize our notion of coding/decoding just a little to allow decoding functions which take *two* code numbers c and d, and an index number i, as follows:

A three-place decoding function is a function $decode(c, d, i)$ such that, for *any* finite sequence of natural numbers $k_0, k_1, k_2, \ldots, k_n$ there is a *pair* of code numbers c, d such that, for every $i \leq n$, $decode(c, d, i) = k_i$.

A three-place decoding-function will obviously do just as well as a two-place function to help us express facts about finite sequences.

Even with this liberalization, though, it still isn't immediately obvious how to define a decoding function in terms of the functions built into basic arithmetic. But Gödel neatly solved the problem with his β-function. Put

$$\beta(c, d, i) =_{\text{def}} \text{ the remainder left when } c \text{ is divided by } d(i + 1) + 1.$$

Then we have

Theorem 25. *For any finite sequence of numbers $k_0, k_1, \ldots, k_n$, we can find a suitable pair of numbers c, d such that for $i \leq n$, $\beta(c, d, i) = k_i$.*

This arithmetical claim should look intrinsically plausible. As we divide c by $d(i + 1) + 1$, then for different values of i ($0 \leq i \leq n$) we'll get a sequence of $n + 1$ remainders. Vary c and d, and the sequence of remainders will vary. The permutations as we vary c and d without limit *appear* to be simply endless. We just need to check, then, that appearances don't deceive, and we *can* always find a (big enough) c and a (smaller) d which makes the sequence of remainders match a given $n + 1$-term sequence of numbers.[1]

(d)   OK: we said that $y = x!$ just in case

1. There is a sequence of numbers $k_0, k_1, \ldots, k_x$ such that: $k_0 = 1$, and if $i < x$ then $k_{Si} = k_i \times Si$, and $k_x = y$.

And we now know, thanks to Gödel, that we can reformulate this as follows:

2′. There is some pair of code numbers c, d such that: $\beta(c, d, 0) = 1$, and if $i < x$ then $\beta(c, d, Si) = \beta(c, d, i) \times Si$, and $\beta(c, d, x) = y$.

But the β-function is defined in terms of the elementary arithmetic notion of remainder-on-division. So we'll expect that it can be expressed in L_A by some open wff which we will abbreviate B (so for particular numbers a, b, j, k, $\mathsf{B}(\bar{\mathsf{a}}, \bar{\mathsf{b}}, \bar{\mathsf{j}}, \bar{\mathsf{k}})$ is true if and only if $\beta(a, b, j) = k$).

Assuming a suitable wff B is available, (2′) goes into L_A as follows:

3. $\exists \mathsf{c} \exists \mathsf{d} \{ \mathsf{B}(\mathsf{c}, \mathsf{d}, 0, \mathsf{S}0) \wedge$
   $(\forall \mathsf{i} \leq \mathsf{x})[\mathsf{i} \neq \mathsf{x} \rightarrow \exists \mathsf{v} \exists \mathsf{w} \{ (\mathsf{B}(\mathsf{c}, \mathsf{d}, \mathsf{i}, \mathsf{v}) \wedge \mathsf{B}(\mathsf{c}, \mathsf{d}, \mathsf{Si}, \mathsf{w})) \wedge \mathsf{w} = \mathsf{v} \times \mathsf{Si} \}] \wedge$
   $\mathsf{B}(\mathsf{c}, \mathsf{d}, \mathsf{x}, \mathsf{y}) \}$.

Abbreviate all that by '$\mathsf{F}(\mathsf{x}, \mathsf{y})$'. Then this expresses (2′) which is equivalent to (1). Hence $\mathsf{F}(\mathsf{x}, \mathsf{y})$ expresses the factorial function. Neat! – assuming we can indeed fill in the remaining details and define B.

(e)   Finally, then, we need to come up with an L_A wff B which expresses β.

Well, reflect that the concept of a remainder on division can be elementarily defined in terms of multiplication and addition. The remainder when a is divided by b equals y, just when there is some number u (no greater than a) such that $a = \{ b \times u \} + y$, where $y < b$.

Similarly, the remainder left when c is divided by $d(i+1)+1$ equals y just when there is some number u (no greater than c) such that $c = \{ (d(i+1)+1) \times u \} + y$, where $y < d(i + 1) + 1$, or equivalently $y \leq d(i + 1)$.

So consider the following open wff (with its initial bounded quantifier):

$$\hat{\mathsf{B}}(\mathsf{c}, \mathsf{d}, \mathsf{i}, \mathsf{y}) =_{\text{def}} (\exists \mathsf{u} \leq \mathsf{c})[\mathsf{c} = \{ \mathsf{S}(\mathsf{d} \times \mathsf{Si}) \times \mathsf{u} \} + \mathsf{y} \wedge \mathsf{y} \leq (\mathsf{d} \times \mathsf{Si})].$$

This, as we wanted, expresses our three-place Gödelian β-function in L_A (for remember, we can define '$\leq$' in L_A).

Job done. Except that, for technical reasons, it turns out to be useful to add to $\hat{\mathsf{B}}$ a clause that reflects that the β function is indeed a function (and hence, in particular, when it sends inputs (c, d, i) to the value y, it can't send those same inputs to any smaller value z). So let's define

$$\mathsf{B}(\mathsf{c}, \mathsf{d}, \mathsf{i}, \mathsf{y}) =_{\text{def}} \hat{\mathsf{B}}(\mathsf{c}, \mathsf{d}, \mathsf{i}, \mathsf{y}) \wedge (\forall \mathsf{z} \leq \mathsf{y})(\mathsf{z} \neq \mathsf{y} \rightarrow \neg \hat{\mathsf{B}}(\mathsf{c}, \mathsf{d}, \mathsf{i}, \mathsf{z}))$$

This then will be our official way of expressing the β function in L_A.

---

[1]Mathematical completists: see *IGT2*, §15.2, fn. 4 for a proof that this works!

9.3 L_A can express all p.r. functions

We now want to show that what goes for the factorial function goes for any p.r. function. In other words, using the β-function trick again, we can now generalize to show that L_A can express any p.r. function. (If you worked through the last section, you are welcome to take this generalizing claim on trust.)

We already know from §8.3 the standard strategy for showing that something is true of all p.r. functions. So suppose that the following three propositions are all true:

E1. L_A can express the initial functions. (See Defn. 31.)

E2. If L_A can express the functions g and h, then it can also express a function f defined by composition from g and h. (See Defn. 30.)

E3. If L_A can express the functions g and h, then it can also express a function f defined by primitive recursion from g and h. (See Defn. 29.)

Then by the argument of §8.3, those assumptions will be enough to establish our desired general result. So how can we prove (E1) to (E3)?

Proof of E1. Just look at cases. The successor function $Sx = y$ is of course expressed by the open wff $Sx = y$.

The zero function, $Z(x) = 0$ is expressed by the wff $Z(x, y) =_{\text{def}} (x = x \land y = 0)$.

Finally, the three-place function $I_2^3(x, y, z) = y$, to take just one example of an identity function, is expressed by the wff $I_2^3(x, y, z, u) =_{\text{def}} y = u$ (or we could use $(x = x \land y = u \land z = z)$ if we'd like x and z actually to appear in the wff). Likewise for all the other identity functions.   ☒

Proof of E2. Suppose, to take a simple example, that g and h are one-place functions, expressed by the wffs $G(x, y)$ and $H(x, y)$ respectively. Then, the function $f(x) = h(g(x))$ is evidently expressed by the wff $\exists z(G(x, z) \land H(z, y))$.

For suppose $g(m) = k$ and $h(k) = n$, so $f(m) = n$. Then by hypothesis $G(\overline{m}, \overline{k})$ and $H(\overline{k}, \overline{n})$ will be true, and hence $\exists z(G(\overline{m}, z) \land H(z, \overline{n}))$ is true, as required. Conversely, suppose $\exists z(G(\overline{m}, z) \land H(z, \overline{n}))$ is true. Then since the quantifiers run over numbers, $(G(\overline{m}, \overline{k}) \land H(\overline{k}, \overline{n}))$ must be true for some k. So we'll have $g(m) = k$ and $h(k) = n$, and hence $f(m) = h(g(m)) = n$ as required.

Other cases – where g and/or h are multi-place functions – can be handled similarly.   ☒

Proof of E3. We need to show that we can use the β-function trick again and – exactly following the model of our treatment of the factorial – prove more generally that, if the function f is defined by recursion from functions g and h which are already expressible in L_A, then f is also expressible in L_A.

We are assuming that

$$f(\vec{x}, 0) = g(\vec{x})$$
$$f(\vec{x}, Sy) = h(\vec{x}, y, f(\vec{x}, y)).$$

(Remember, $\vec{x}$ indicates some variables being 'carried along for the ride', that don't change in the course of the recursion that defines $f(\vec{x}, Sy)$ in terms of $f(\vec{x}, y)$.)

This definition amounts to fixing the value of $f(\vec{x}, y) = z$ thus:

1. There is a sequence of numbers $k_0, k_1, \ldots, k_y$ such that: $k_0 = g(\vec{x})$, and if $i < y$ then $k_{Is} = h(\vec{x}, u, k_i)$, and $k_y = z$.

Equivalently, using our β-function again, $f(\vec{x}, y) = z$ is true when:

2. There is some c, d, such that: $\beta(c, d, 0) = g(\vec{x})$, and if $i < y$ then $\beta(c, d, Si) = h(\vec{x}, i, \beta(c, d, i))$, and $\beta(c, d, y) = z$.

Suppose we can already express the n-place function g by a $(n + 1)$-variable expression G, and the $(n+2)$-variable function h by the $(n+3)$-variable expression H. Then – using '$\vec{x}$' to indicate a suitable sequence of n variables – (2) can be rendered into L_A by

3. $\exists c \exists d \{\exists k[B(c, d, 0, k) \wedge G(\vec{x}, k)] \wedge$
   $(\forall i \leq y)[i \neq y \rightarrow \exists v \exists w \{(B(c, d, i, v) \wedge B(c, d, Si, w)) \wedge H(\vec{x}, i, v, w)\}] \wedge$
   $B(c, d, y, z)\}$.

Abbreviate this defined wff as $\varphi(\vec{x}, y, z)$; it is then evident that φ will serve to express the p.r. defined function f. Which gives us the desired result E3. ⊠

So, we've shown how to establish each of the claims E1, E2 and E3. Hence

Theorem 26. *The language L_A can express all primitive recursive functions.*

Proof. For any p.r. function f, there is a sequence of functions $f_0, f_1, f_2, \ldots, f_k$ where each f_j is either an initial function or is constructed out of previous functions by composition or recursion, and $f_k = f$. Corresponding to that sequence of functions we can write down a sequence of L_A wffs which express each of those functions in turn. We write down the E1 expression corresponding to an initial function. If f_j comes from two previous functions in the sequence by composition, we use an existential quantifier construction as in E2 to write down a wff built out of the wffs expressing the two previous functions. And if f_j comes from two of the previous functions by recursion, we use the β-function trick and write down a (3)-style expression built out of the wffs expressing the two previous functions. So eventually we build up to the wff which expresses f_k, i.e. our target function f. ⊠

9.4 Canonical wffs for expressing p.r. functions are Σ_1

Let's say that

Defn. 36. *An L_A wff canonically expresses the p.r. function f if it recapitulates a definitional chain for f by being constructed in the manner described in the proof of Theorem 26.*

We can express a given p.r. function f by other wffs too (for a start, by adding redundant clauses): but it is the *canonical* ones from which we can read off a full definitional chain for f which will interest us the most.

Now, a canonical wff which reflects a full definition of f is built up starting from wffs expressing initial wffs. Those starter wffs are Δ_0 wffs – see the Proof for E1 – and hence Σ_1.

Suppose g and h are one-place functions, expressed by the Σ_1 wffs $\mathsf{G(x,y)}$ and $\mathsf{H(x,y)}$ respectively. The function $f(x) = h(g(x))$ is expressed by the wff $\exists \mathsf{z(G(x,z) \land H(z,y))}$ – as in the Proof for E2 – which is Σ_1 too. For that is equivalent to a wff with the existential quantifiers pulled from the front of the Σ_1 wffs G and H out to the very front of the new wff. Similarly for other cases of composition.

Finally, look at the case where f is defined from g and h by primitive recursion, where g and h are p.r. functions which can be expressed by Σ_1 wffs. Then f can be expressed by a wff of the kind (3) – as in our proof of E3. And this too is in fact Σ_1. For a bit of manipulation will show that B is Δ_0: and a little more manipulation will show that (3) is equivalent to what we get when we drag all the existential quantifiers buried at the front of each of B, G and H to the very front of the wff.[2]

So our recipe for building a canonical wff stage by stage takes us from Σ_1 wffs to Σ_1 wffs. Which yields the stronger

Theorem 27. *L_A can express any p.r. function f by a Σ_1 wff which recapitulates a full definitional chain for f.*

9.5   Q can capture all p.r. functions

We have shown that the language of the theory Q can *express* all p.r. functions using Σ_1 wffs. We now want to show that this theory (and hence any stronger one) can also *capture* all those functions using Σ_1 wffs.

But hold on! We haven't yet said what it is for a theory to capture a function. So we first need to explain that. Recall our earlier account of what it is to capture a property or relation. In particular, recall

Defn 18. *The theory T captures the two-place numerical relation R by the open wff $\varphi(\mathsf{x,y})$ iff, for any m, n,*
    i. if m has the relation R to n, then $T \vdash \varphi(\overline{\mathsf{m}}, \overline{\mathsf{n}})$,
    ii. if m does not have the relation R to n, then $T \vdash \neg\varphi((\overline{\mathsf{m}}, \overline{\mathsf{n}})$.

Thinking of $f(m) = n$ as stating a two-place relation between m and n we might expect the definition for capturing a function to have the same shape:

Defn. 37. *The theory T captures the one-place function f by the open wff $\psi(\mathsf{x,y})$ iff, for any m, n,*

---

[2]Fine print: Yes, we have to drag existentials to the front of a wff past a universal, and that is usually wicked! – however the universal here is *bounded*, so is 'really' just a tame conjunction, and this does allow us to get the existentials all at the front.

    i. if $f(m) = n$, then $T \vdash \psi(\overline{m}, \overline{n})$,
    ii. if $f(m) \neq n$, then $T \vdash \neg\psi(\overline{m}, \overline{n})$.

But (for technical reasons we are not going to fuss about here) it turns out to be useful to add a further requirement

    iii. $T \vdash \exists! y \psi(\overline{m}, y)$,

where '$\exists!$' is the uniqueness quantifier, 'there exists exactly one'. In other words, T also 'knows' that ψ is functional (associates a number to just one value).

Our target, then, is the following theorem:

Theorem 28. *The theory* Q *can capture any p.r. function by a Σ_1 wff.*

Proof outline. The conceptually easiest route is to use again the same overall strategy we used in proving that Q's language can express every p.r. function. Suppose then that we can prove

C1. Q can capture the initial functions.

C2. If Q can capture the functions g and h, then it can also capture a function f defined by composition from g and h.

C3. If Q can capture the functions g and h, then it can also capture a function f defined by primitive recursion from g and h.   ⊠

Then it follows that Q can capture any p.r. function.

So how do we prove C1? We just check that the formulas which we said in §9.3 *express* the initial functions in fact serve to *capture* the initial functions in Q.

How do we prove C2? Suppose g and h are one-place functions, captured by the wffs $G(x, y)$ and $H(x, y)$ respectively. Then we prove that the function $f(x) = h(g(x))$ is captured by the wff $\exists z(G(x, z) \wedge H(z, y))$. We can generalize the result.

And how do we prove C3? This is the tedious case that takes hard work! We need to show that B not only expresses but captures Gödel's β-function.[3] And then we use that fact to prove that if the n-place function g is captured by a $(n+1)$-variable expression G, and the $(n+2)$-variable function h by the $(n+3)$-variable expression H, then a wff built to the pattern of (3) in §9.3 captures the function f defined by primitive recursion from g and h. Not surprisingly, details get messy.[4]

So take a definitional chain for defining a p.r. function. Follow the step-by-step instructions implicit in §9.3 about how to build up a wff which in effect recapitulates that recipe. You'll get a wff that not only canonically expresses but captures the function in Q (and so captures it too in any stronger theory

---

[3]See again the very end of §9.2. This is where the dodge of using using B rather than $\hat{\text{B}}$ comes into its own; the added clause in B is there to reflect the functional character of the β-function.

[4]Not difficult, but messy – as you can see from *IFL2*, particularly §17.3.

which contains the language of basic arithmetic). And we have already seen that that the canonical wff in question will be Σ_1.[5]

9.6 Expressing/capturing properties and relations

Just a brief coda, linking what we've done in this chapter with the last section of the previous chapter.

We said in Defn. 34 that the characteristic function c_P of a monadic numerical property P is defined by setting $c_P(m) = 0$ if m is P and $c_P(m) = 1$ otherwise. And a property P is said to be p.r. decidable if its characteristic function is p.r.

Now, suppose that P is p.r.; then c_P is a p.r. function. So L_A can express c_P by a two-place Σ_1 wff $\mathsf{c}_P(\mathsf{x}, \mathsf{y})$. So if m is P, i.e. $c_P(m) = 0$, then $\mathsf{c}_P(\overline{\mathsf{m}}, 0)$ is true. And if m is not P, i.e. $c_P(m) \neq 0$, then $\mathsf{c}_P(\overline{\mathsf{m}}, 0)$ is not true. Hence, by the definition of expressing-a-property, the wff $\mathsf{c}_P(\mathsf{x}, 0)$ serves to express the p.r. property P. The point generalizes from monadic properties to many-place relations. So as an easy corollary of Theorem 27 we get:

Theorem 29. L_A *can express all p.r. decidable properties and relations, again using Σ_1 wffs.*

Similarly, suppose again that the monadic property P is p.r. so c_P is a p.r. function. Then Q can capture c_P by a two-place Σ_1 wff $\mathsf{c}_P(\mathsf{x}, \mathsf{y})$. So if m is P, i.e. $c_P(m) = 0$, then $Q \vdash \mathsf{c}_P(\overline{\mathsf{m}}, 0)$. And if m is not P, i.e. $c_P(m) \neq 0$, then $Q \vdash \neg\mathsf{c}_P(\overline{\mathsf{m}}, 0)$. Hence, by the definition of capturing-a-property, the wff $\mathsf{c}_P(\mathsf{x}, 0)$ serves to capture the p.r. property P in Q. The point trivially generalizes from monadic properties to many-place relations. So as an easy corollary of Theorem 28 we get:

Theorem 30. Q *can capture all p.r. decidable properties and relations, again using Σ_1 wffs.*

---

[5] The important distinction between those ways of formally expressing/capturing a p.r. function which directly recapitulate a definitional chain and other ways of doing the job is standard; the use of the label 'canonical' to mark the distinction doesn't seem to be so widespread.

10 The arithmetization of syntax

Back in Chapter 3, we outlined how Gödel proved his First Incompleteness Theorem. One idea we introduced then was the arithmetization of syntax. We return now to investigate this pivotal idea.

10.1 Gödel-numbering

It was Hilbert[1] who first emphasized that the syntactic objects that comprise formal theories (the wffs, the proofs) are *finite* objects, and so we only need a mathematics of finite objects to deal with the syntactic properties of theories. We'll return to discuss the significance that this insight had for Hilbert in Chapter 17. But for now, we want Gödel's great twist on this idea: when we are dealing with finite objects, we can give them numerical codes. Hence we can use *arithmetic* to talk – via the coding – about syntactic properties of theories (including the syntactic properties of theories of arithmetic in particular).

Let's start by concentrating on the particular case of coding up expressions of the language L_A. There are various ways of doing the coding. We might use a version of the sort of coding we met in §3.3. But here we'll use the general style of coding used by Gödel himself. Nothing really hangs on the choice: any coding scheme will do so long as it is 'normal' in a sense which we'll explain shortly.

Suppose, then, that our version of L_A has the usual logical symbols (connectives, quantifiers, identity, brackets), and symbols for zero and for the successor, addition and multiplication functions: associate all those symbols with odd numbers (different symbol, different code number, of course). L_A, as a standard first-order language, also has the usual inexhaustible supply of variables, which we'll associate with even numbers. So, to pin that down, let's fix on this preliminary series of *basic codes*:

$$\neg \ \wedge \ \vee \ \rightarrow \ \leftrightarrow \ \forall \ \exists \ = \ (\ ) \ 0 \ S \ + \ \times \ x \ y \ z \ \ldots$$
$$1 \ \ 3 \ \ 5 \ \ 7 \ \ \ 9 \ \ 11 \ \ 13 \ \ 15 \ \ 17 \ \ 19 \ \ 21 \ \ 23 \ \ 25 \ \ 27 \ \ 2 \ \ 4 \ \ 6 \ \ \ldots$$

Our Gödelian numbering scheme for expressions is now defined in terms of this table of basic codes as follows:

---

[1]David Hilbert (1862–1943) was one of the most influential and wide-ranging mathematicians of the late nineteenth and early twentieth century. Working with his research assistant Paul Bernays (1888–1977), Hilbert's influence on the development of logic was profound.

Defn. 38. *Suppose that the expression e is the sequence of k symbols and/or variables $s_1, s_2, \ldots, s_k$. Then e's Gödel number (g.n.) is calculated by taking the basic code-number c_i for each s_i in turn, using c_i as an exponent for the i-th prime number π_i, and then multiplying the results, to get $2^{c_1} \cdot 3^{c_2} \cdot 5^{c_3} \cdot \ldots \cdot \pi_k^{c_k}$.*

For example:

   i. The single symbol 'S' has the g.n. 2^{23} (the first prime raised to the appropriate power as read off from our correlation table of basic codes).

   ii. The standard numeral SS0 has the g.n. $2^{23} \cdot 3^{23} \cdot 5^{21}$ (the product of the first three primes raised to the appropriate powers).

  iii. The wff

$$\exists y\,(S0 + y) = SS0$$

has the g.n.

$$2^{13} \cdot 3^4 \cdot 5^{17} \cdot 7^{23} \cdot 11^{21} \cdot 13^{25} \cdot 17^4 \cdot 19^{19} \cdot 23^{15} \cdot 29^{23} \cdot 31^{23} \cdot 37^{21}.$$

That last number is, of course, *enormous*. So when we say that it is elementary to decode the resulting g.n. by taking the exponents of prime factors, we don't mean that the computation is quick. We mean that the computational routine required for the task – namely, repeatedly extracting prime factors – involves no more than the mechanical operations of elementary arithmetic. And of course, because numbers are uniquely decomposable into prime factors, the decoding of any number will be unique – either a particular expression or a null result.

Now, as well as talking about individual wffs via their code numbers, we will also want to talk about whole proofs via *their* code numbers. But how *do* we code for proof-arrays?

The details will obviously depend on the kind of proof system we adopt for the theory we are using. Suppose though, for simplicity, we consider theories with a Hilbert-style axiomatic system of logic. In this rather old-fashioned framework, proof-arrays are simply *linear sequences* of wffs. A nice way of coding these sequences is by what we'll call *super Gödel numbers*.

Defn. 39. *Given a sequence of wffs or other expressions $e_1, e_2, \ldots, e_n$, we first code each e_i by a regular g.n. g_i, to yield a corresponding sequence of numbers $g_1, g_2, \ldots, g_n$. We then encode this sequence of regular Gödel numbers using a single super g.n. by repeating the trick of multiplying powers of primes to get $2^{g_1} \cdot 3^{g_2} \cdot 5^{g_3} \cdot \ldots \cdot \pi_n^{g_n}$.*

Decoding a super g.n. therefore involves two steps of taking prime factors: first find the sequence of exponents of the prime factors of the super g.n.; then treat each of those exponents in turn as a regular g.n., and take prime factors again to arrive back at a sequence of expressions.

10.2   The arithmetization of syntactic properties/relations

In this section, we will continue to focus on the language L_A and on the particular theory PA which is built in that language. But as we will stress in the next section, similar results will apply mutatis mutandis to any sensibly axiomatized formal theory.

Recall Defn. 15 from §3.4. In the present context, this becomes:

Defn. 40. *Given our scheme for Gödel-numbering* PA *expressions and sequences of expressions, we can define the following properties/relations:*

> $Wff(n)$ *iff* n *is the Gödel number of a* L_A *wff.*
> $Sent(n)$ *iff* n *is the Gödel number of a* L_A *sentence.*
> $Prf(m, n)$ *iff* m *is the super Gödel number of a* PA-*proof of the* L_A
> *sentence with code number* n.

Then we have the following key result:

Theorem 31. *Wff and Sent are p.r. decidable properties; and Prf is a p.r. decidable relation.*

How do we show this? Writing at the very beginning of the period when concepts of computation were being forged, Gödel couldn't expect his audience to take anything on trust about what was or wasn't '*rekursiv*' or – as we would now put it – primitive recursive. He therefore had to do all the hard work of explicitly showing how to define such properties (or their characteristic functions) by a long chain of definitions by composition and recursion.

However, assuming only a modest familiarity with the ideas of computer programs and p.r. functions, and accepting Theorem 23, we can perhaps short-cut all that effort and be persuaded by the following:[2]

Informal proof idea. To determine whether $Wff(n)$, first decode n: a simple algorithm does the job, one doesn't require an open-ended search. Now ask: is the resulting expression a wff of the language L_A? That's algorithmically decidable – and again no open-ended search is required: the length of the required computation will be fixed by the length of the expression. So neither stage of the decision procedure will involve any open-ended search; they can be done by programs using just bounded-in-advance 'for' loops.

The second stage of this decision procedure works on simple strings of symbols. But we can imagine a parallel procedure, operating directly on numbers coding for symbol-strings. So we can think of the whole computation as done on numbers, still without open-ended searches (its only looping structures are bounded-in-advance 'for' loops); so Wff is p.r. decidable. Similarly for $Sent(n)$

To determine whether $Prf(m, n)$, first doubly decode the super Gödel number m: that's a mechanical exercise. Now ask: is the result a sequence of PA

---

[2]Frankly, we *are* cutting corners here. If you aren't sufficiently persuaded, then be assured that we *can* prove Theorem 31 the hard way, as explained in *IGT2*.

wffs? That's algorithmically decidable (since it is decidable whether each separate string of symbols is a wff). If it does decode into a sequence of wffs, ask next: is this sequence a properly constructed PA proof? That's decidable too (check whether each wff in the sequence is either an axiom or is an immediate consequence of previous wffs by one of the rules of inference of PA's Hilbert-style logical system). If the sequence is a proof, ask: does its final wff have the g.n. n? That's again decidable. Finally, assuming theorems have to be sentences, ask whether $Sent(n)$ is true.

Putting things together, there is a computational procedure for telling whether $Prf(m, n)$ holds. Moreover, at each stage, the computation involved is once more a straightforward, bounded procedure without any open-ended searches. The length of the sequence of expressions with code m puts a ceiling on the work we have to do. So the procedure is one that can be written up as a program using only bounded 'for'-loops (and we could in principle do all the computations on numbers rather than symbols). Hence Prf is also p.r. decidable. ⊠

10.3 Generalizing

Using a subscript to highlight the particular theory we are considering, Theorem 31 tells us that the key relation Prf_{PA} defined using one particular Gödel-numbering scheme is primitive recursive.

However, our adopted numbering scheme was fairly arbitrarily chosen. We could, for example, shuffle around the preliminary assignment of basic codes to get a different Gödel-style numbering scheme; or we could use a scheme that isn't based on powers of primes. So could it be that a relation like Prf_{PA} is p.r. when defined in terms of our particular numbering scheme and not p.r. when defined in terms of some alternative but equally acceptable scheme?

Well, what counts as 'acceptable' here? The key feature of our Gödelian scheme is this: there is a pair of algorithms, one of which takes us from an L_A expression to its code number, the other of which takes us back again from the code number to the original expression. Moreover, in following through these algorithms, the upper length of the computation is determined by the length of the L_A expression to be encoded or the size of the number to be decoded: *we don't have to go on unbounded searches*. The computations can be done just using bounded 'for' loops.

And now let's generalize this idea. We will say:

Defn. 41. *A normal Gödel-numbering scheme is one which deploys coding and decoding algorithms which don't involve any open-ended searches.*

Back in Defn. 14, we required the coding and decoding procedures in a Gödel-numbering scheme to be *effective* procedures. We are now tightening this requirement. From now on, we assume that the required effective procedures can be done without open-ended searches, so that our coding schemes are normal. And with this assumption in place, we can simply remark that, whichever (normal) coding scheme for PA we choose, the informal proof we gave will go through as before: Prf_{PA}, now redefined using the new scheme, will still be p.r. decidable.

What about Prf_T for other theories T (assuming a normal Gödel-numbering scheme is in play)? In other words, take the relation that holds between m and n when (according to our scheme) m codes for a T-proof of the sentence coded by n; is this a p.r. relation again for other theories T?

Suppose T is not just effectively axiomatized, but is put together so that we can mechanically check whether a purported T-proof is a kosher proof *without* having to set out on an unbounded search (the length of a candidate proof sets a limit known in advance to the number of steps it will take to check whether it *is* a proof). In other words, suppose that checking a proof can be done by a procedure that can be regimented using nothing more exotic than 'for' loops. Then, by the same informal proof as before, the relation Prf_T which holds when m numbers a proof of the wff with number n will still be primitive recursive.

Let's say that

Defn. 42. *A theory T is* p.r. axiomatized *when it is so axiomatized as to make the relation Prf_T (defined using a normal Gödel-numbering scheme) a primitive recursive relation.*

Then *any* usual kind of formal theory you dream up will actually be p.r. axiomatized if it is effectively axiomatized at all. We *never* in practice formalize a theory in such a way that (i) it is effectively axiomatized, but (ii) we can only effectively check whether a given array of expressions is a well-constructed proof by some unbounded search(es). Hold onto that important observation!

10.4   Some cute notation

We now introduce a rather pretty bit of notation. Assume we have chosen some system for Gödel-numbering the expressions of a language L. Then

Defn. 43. *If φ is an L-expression, then we'll use '$\ulcorner\varphi\urcorner$' in our logicians' augmented English to denote φ's Gödel number. And we use '$\overline{\ulcorner\varphi\urcorner}$' as an abbreviation in our formal arithmetic for the standard numeral for the number $\ulcorner\varphi\urcorner$.*[3]

Borrowing corner quotes for this new use is quite appropriate because the number $\ulcorner\varphi\urcorner$ can be thought of as referring to the expression φ via our coding scheme. (Sometimes, we'll write the likes of $\ulcorner\mathsf{U}\urcorner$ where U abbreviates an L_A wff: we mean here, of course, the Gödel-number for the unabbreviated original wff that U stands in for.)

---

[3]In *IGT2*, I use the corner-bracket notation to stand both for Gödel numbers *and* (still without overlining) for the standard numerals for Gödel numbers, letting context disambiguate. That too is a common convention. But in these notes, in the interest of maximal clarity – if only as a helpful ladder that you can throw away once climbed! – I *will* here use the clumsier overlining notation for numerals for Gödel numbers.

And given we used the overlined expression '$\bar{n}$' to abbreviate the standard numeral for the number n, it is quite natural to use the same convention again in using the overlined '$\overline{\ulcorner\varphi\urcorner}$' to abbreviate the standard numeral for $\ulcorner\varphi\urcorner$. A simple example to illustrate:

1. 'SS0' is an L_A expression, the standard numeral for 2.

2. On our numbering scheme $\ulcorner SS0 \urcorner$, the g.n. of 'SS0', is $2^{23} \cdot 3^{23} \cdot 5^{21}$.

3. Then '$\overline{\ulcorner SS0 \urcorner}$' is shorthand, used as an abbreviation for the standard L_A-numeral for that g.n., i.e. as an abbreviation for 'SSS . . . S0' with $2^{23} \cdot 3^{23} \cdot 5^{21}$ occurrences of 'S'!

10.5 Diagonalization

Finally, we use our new notation in defining a simple construction which will play an all-important role in the coming chapters (so read carefully!). Assume we are working with a language with standard numerals, and that we have a Gödel-numbering scheme in play: then

Defn. 44. *The diagonalization of a wff φ with one free variable is the wff $\varphi(\overline{\ulcorner\varphi\urcorner})$*

Let's clarify: now making the free variable explicit, the diagonalization of the wff $\varphi(x)$ is what you get by substituting for its free variable x the numeral for the Gödel number of the whole wff $\varphi(x)$. For example, the diagonalization of $\exists y F(x, y)$ is $\exists y F(\overline{\ulcorner \exists y F(x, y) \urcorner}, y)$.

Why is this substitution operation called *diagonalization*? Well compare the 'diagonal' construction we encountered in §4.3. There, we counted off wffs $\varphi_0(x)$, $\varphi_1(x)$, $\varphi_2(x)$ . . . in an enumeration of wffs with one free variable; and then we substituted (the numeral for) the index d for the free variable in the wff φ_d, to form $\varphi_d(\bar{d})$. We can now think of the Gödel number of a wff with one free variable as another way of indexing that wff in a list of such wffs. And so, in our new diagonal construction, we are again substituting (the numeral for) the index of a wff for the free variable in the wff.

Diagonalization is an elementary mechanical operation on expressions. Hence, assuming that we are working with a normal Gödel numbering scheme for the language L, we can expect this to be true:

Theorem 32. *There is a p.r. function $diag(n)$ which, when applied to a number n which is the g.n. of some L-wff with one free variable, yields the g.n. of that wff's diagonalization, and yields 0 otherwise.*

Proof. Try treating n as a g.n., and seek to decode it. If you don't get an expression with one free variable, return 0. Otherwise you get a wff of the type $\varphi(x)$, and can then form the wff $\varphi(\bar{n})$, which is its diagonalization (since by assumption n is its g.n.). Then we work out the g.n. of this resulting wff to compute $diag(n)$.

This procedure doesn't involve any unbounded searches. So we will be able to program the procedure using just 'for' loops. Hence *diag* is a p.r. function. ⊠

11 The First Incompleteness Theorem, semantic version

Let's quickly review some crucial ideas that we have met in the last couple of chapters:

i. We fixed on a particular scheme for coding up wffs of PA's language L_A by using Gödel numbers ('g.n.' for short), and for coding up PA-proofs by super Gödel numbers (assuming for convenience that these proofs are simple sequences of wffs). On our scheme, the algorithms which take us from expressions to code numbers and back again don't involve any open-ended searches. Call a coding scheme with this feature *normal*. (§10.2)

ii. $Prf(m, n)$ is the relation which holds if and only if m is the super g.n. (on our normal scheme) of a sequence of wffs that is a PA proof of a sentence with g.n. n. This relation is primitive recursive. (§10.2)

iii. Any p.r. function or relation can be *expressed* by a wff of PA's language L_A. In particular, we can choose a Σ_1 wff Prf which *canonically* expresses *Prf* by recapitulating the definitional chain for the relation's characteristic function. (§§3.5, 9.3)

iv. Any p.r. function or relation can be *captured* in Q and hence in PA. For example, *Prf* can be captured by a Σ_1 wff (again one which recapitulates the relation's p.r. definition). (§9.5)

v. Notation: If φ is an expression, then we'll denote its Gödel number in our logician's English by '$\ulcorner\varphi\urcorner$'. We use '$\overline{\ulcorner\varphi\urcorner}$' as an abbreviation for the standard numeral for $\ulcorner\varphi\urcorner$. (§10.4)

vi. The diagonalization of a wff φ with one free variable is $\varphi(\overline{\ulcorner\varphi\urcorner})$. The function *diag* sends the g.n. of a wff to the g.n. of its diagonalization and is primitive recursive. (§10.5)

For what follows, it isn't necessary that you know the *proofs* of the claims we've just summarized: but do pause to check that you at least fully understand what the various claims *mean*.

11 The First Incompleteness Theorem, semantic version

11.1 Constructing a Gödel sentence

(a)   As we announced right back in Chapter 3, Gödel is going to tell us how to construct a wff G in PA that is true if and only if it is unprovable in PA. We now have an inkling of how he can do that. Arithmetization allows us to construct wffs which express proof-relations like *Prf*. Moreover, wffs can contain numerals which refer to numbers which – via Gödel coding – are in turn correlated with wffs. Maybe, if we are cunning, we can get a wff to be 'about' itself.

At the end of the last chapter we met the basic construction we are going to use, namely diagonalization. Recall, then, that the relation $Prf(m, n)$ is defined to hold just when m is the super g.n. for a PA proof of the wff with g.n. n. Now for a simple tweak:[1]

Defn. 45. *The relation* $Prfd(m, n)$ *holds if and only if* m *is the super g.n. for a* PA *proof of the diagonalization of the wff with g.n.* n.

Suppose n numbers a wff $\varphi(\mathsf{x})$; then $Prfd(m, n)$ holds if m numbers a proof of the wff $\varphi(\ulcorner \mathsf{n} \urcorner)$ (which is the wff with g.n. $diag(n)$).

Theorem 33. *Prfd is primitive recursive.*

Official proof. $Prfd(m, n)$ is true iff $Prf(m, diag(n))$. Let c_{Prf} be the characteristic function of Prf, which is primitive recursive. Then the characteristic function of $Prfd$ is $c_{Prfd}(x, y) =_{\text{def}} c_{Prf}(x, diag(y))$, so – being a composition of p.r. functions – c_{Prfd} is primitive recursive too. $\boxtimes$

Informal proof. We just remark that, as with $Prf(m, n)$, we can mechanically check whether $Prfd(m, n)$. Just decode m. Check whether it gives a sequence of wffs. If it does, check whether it is a PA proof. If it is, check whether the concluding wff of the proof is the result of taking an open wff whose g.n. is n and diagonalizing that wff. That involves no open-ended searches. An algorithm involving only 'for' loops will suffice. So *Prfd* is primitive recursive. $\boxtimes$

Since *Prfd* is p.r. it can be both expressed and captured in PA by a canonical Σ_1 wff, i.e. one which recapitulates a definition of the (characteristic function) of the p.r. relation. So let's say

Defn. 46. $\mathsf{Prfd(x, y)}$ *stands in for some* Σ_1 *wff which canonically expresses and captures Prfd.*

(b)   And now comes the simple but ingenious Gödelian construction! First we form the open wff we'll abbreviate as U (think 'unprovable'), or to make its free variable explicit,

Defn. 47. $\mathsf{U(y)} =_{\text{def}} \neg \exists \mathsf{x}\, \mathsf{Prfd(x, y)}$.

---

[1] In *IGT2*, I used '*Gdl*' in honour of Gödel rather than '*Prfd*'; the revised notation I use here should be more memorable, since it indicates more clearly the relation which is in question.

Now we diagonalize U, to give

Defn. 48. $G =_{def} U(\ulcorner U \urcorner) = \neg \exists x \, Prfd(x, \ulcorner U \urcorner)$.

And here is the wonderful result:

Theorem 34. G *is true if and only if it is unprovable in* PA.

Proof. Consider what it takes for G to be true (on the interpretation built into L_A), given that the formal predicate Prfd expresses the numerical relation *Prfd*.

G, i.e. $\neg \exists x \, Prfd(x, \ulcorner U \urcorner)$, is true if and only if no number m it is such that $Prfd(m, \ulcorner U \urcorner)$. That is to say, given the definition of *Prfd*, G is true if and only if there is no number m such that m is the code number for a PA proof of the diagonalization of the wff with g.n. $\ulcorner U \urcorner$. But the wff with g.n. $\ulcorner U \urcorner$ is of course U; and its diagonalization is G.

So, G is true if and only if there is no number m such that m is the code number for a PA proof of G. But if G is provable in PA, some number would be the code number of a proof of it. Hence G is true if and only if it is unprovable in PA. $\boxtimes$

Pause and admire this elegant construction!

G – meaning of course the L_A sentence you get when you unpack the abbreviations! – is thus our promised Gödel sentence for PA. We might call it a *canonical* Gödel sentence for three reasons: (a) it is defined in terms of a wff that we said canonically expresses/captures *Prfd*, and (b) because it is the sort of sentence that Gödel himself constructed, so (c) it is the kind of sentence people standardly have in mind when they talk of '*the*' Gödel sentence for PA.

It is true that G will be horribly long when spelt out in quite unabbreviated L_A with its uneconomical standard numerals. But in another way, it is relatively simple. We have the easy result that

Theorem 35. G *is* Π_1.

Proof. $Prfd(x, y)$ is Σ_1. So $Prfd(x, \ulcorner U \urcorner)$ is still Σ_1 (we've just filled up one slot in the open wff with a numeral). So is its existential quantification $\exists x \, Prfd(x, \ulcorner U \urcorner)$ (since the result of adding another existential quantifier to the front of a Σ_1 wff is still Σ_1). Negating to get G gives us a Π_1 wff (since the negation of a Σ_1 wff is Π_1). $\boxtimes$

11.2   What G says

It is often claimed that a canonical Gödel sentence like G actually *says* of itself that it is unprovable. However, that can't be strictly true.

G (when unpacked) is just another sentence of PA's language L_A, the language of basic arithmetic. It is a long wff involving the first-order quantifiers, the connectives, the identity symbol, and 'S', '+' and '×', which all have the standard interpretation built into L_A. In particular, the standard numeral in G refers to a number, not a wff; and the quantifier in G runs over numbers. So G strictly speaking says something about *numbers*, not about wffs and their unprovability.

However there is perhaps a reasonable sense in which G *can* be described as *indirectly* saying that it is unprovable. Note, this is *not* to make play with some radical re-interpretation of G's symbols (for doing *that* would just make any claim about what G says boringly trivial: if we are allowed radical re-interpretations – like spies choosing to borrow ordinary words for use in a secret code – then any string of symbols can be made to say anything). No, it is because the symbols are still being given their *standard* interpretation that we can recognize that the canonically constructed Prfd (when unpacked) will express *Prfd*, given the background framework of Gödel numbering which is involved in the definition of the relation *Prfd*. Therefore, given that coding scheme, we can recognize just from its construction that G will be true when no number m is such that $Prfd(m, \ulcorner U \urcorner)$, and so no number codes for a proof of G. In short, given the coding scheme, we can see just from the way it is constructed that G is true just when it is unprovable. *That* is the limited sense in which, via our Gödel coding, the canonical G signifies or 'indirectly says' that it is unprovable.

11.3   The First Theorem for PA – the semantic version

We already announced in §3.6 that Gödel tells us how to construct a wff which is true if and only if unprovable.[2] And as we showed then, the argument to an incompleteness theorem is now very straightforward. Here it is again.

Assume PA is sound, i.e. proves no falsehoods (because its axioms are true and its logic is truth-preserving). If G could be proved in PA, then PA *would* prove a false theorem (since G is true iff it is *not* provable). That would contradict our supposition that PA is sound. Hence, G is not provable in PA.

But that shows that G *is* true. So ¬G must be false. Hence ¬G cannot be proved in PA either, assuming PA is sound. In Gödel's words, then, G is a 'formally undecidable' sentence of PA (see Defn. 7):

Theorem 36. *If* PA *is sound, then there is a true* Π_1 *sentence* G *such that* PA $\nvdash$ G *and* PA $\nvdash$ ¬G, *so* PA *is negation incomplete.*

If we are happy with the semantic assumption that PA's axioms *are* true on interpretation and so PA *is* sound, the argument for incompleteness is as simple as that – or at least, it's that simple once we have constructed G.

11.4   Generalizing the proof

This line of proof now generalizes. Suppose T is any theory which (i) contains the language of basic arithmetic (see Defn. 11), so T can form standard numerals,

---

[2]If you have been paying close attention to details, you'll spot that we haven't yet *quite* joined up the treatment here with that earlier sketch. But we get all the way there in §13.4.

and we can form the diagonalization of a T-wff with one free variable. Suppose (ii) that we are working with a normal system of Gödel-numbering for T-wffs and T-proofs. Suppose also (iii) that T is p.r. axiomatized in the sense of Defn. 42. Then we can again define a relation $Prfd_T(m, n)$ which holds when m numbers (on our new scheme) a T-proof of the diagonalization of the wff with number n; and this relation will be primitive recursive again.

Continuing to suppose that T's language includes the language of basic arithmetic, T will be able to express the p.r. relation $Prfd_T$ by a Σ_1 wff Prfd_T. Then, just as we did for PA, we'll be able to construct the corresponding Π_1 wff G_T which is true if and only if it is not provable in T. And by exactly the same argument as before we can show, more generally,

Theorem 37. *If T is a sound p.r. axiomatized theory whose language contains the language of basic arithmetic, then there will be a true Π_1 sentence G_T such that $T \nvdash \mathsf{G}_T$ and $T \nvdash \neg\mathsf{G}_T$, so T is negation incomplete.*

Which is at last our first, '*semantic*', version of the generalized First Incompleteness Theorem!

Let's note an immediate corollary:

Theorem 38. *There is no p.r. axiomatized theory framed in the language of L_A whose theorems are all and only the truths of L_A.*

For if the theorems are all true, our theory is sound, and hence it can't be complete.

11.5 Our Incompleteness Theorem is better called an *incompletability* theorem

Here, we just repeat the argument of §2.3: but the point is crucial enough to bear repetition. Suppose T is a sound p.r. axiomatized theory which can express claims of basic arithmetic. Then by Theorem 37 we can find a true G_T such that $T \nvdash \mathsf{G}_T$ and $T \nvdash \neg\mathsf{G}_T$. That *doesn't* mean that G_T is 'absolutely unprovable' (it is *very* unclear whether we can give any good sense to that notion): it just means that G_T is unprovable-in-T.

OK: let's simply augment T by adding G_T as a new axiom, to give the theory $U = T + \mathsf{G}_T$. Then (i) U is still sound (for the old T-axioms are true, the added new axiom is true, and the logic is still truth-preserving). (ii) U is evidently still a p.r. axiomatized theory (why?). (iii) We haven't changed the language. So our Incompleteness Theorem applies, and we can find a sentence G_U such that $U \nvdash \mathsf{G}_U$ and $U \nvdash \neg\mathsf{G}_U$. Since U can prove everything T can prove, that implies $T \nvdash \mathsf{G}_U$ and $T \nvdash \neg\mathsf{G}_U$. In other words, as we put it before, 'repairing the gap' in T by adding G_T as a new axiom leaves some other sentences that are undecidable in T *still* undecidable in the augmented theory.

In sum, our incompleteness theorem tells us that if we keep chucking more and more additional axioms at T, our theory will still remain negation-incomplete, unless it either stops being sound or stops being p.r. axiomatized. In a good sense, T is *incompletable*.

11.6 Comparing old and new semantic incompleteness theorems

Finally in this chapter, let's compare the new Theorem 37 which we *have* proved with the old theorem which we initially announced in §2.2 but of course *didn't* there prove:

Theorem 1. *Suppose T is a formal axiomatized theory whose language contains the language of basic arithmetic. Then, if T is sound, there will be a true sentence G_T of basic arithmetic such that $T \nvdash G_T$ and $T \nvdash \neg G_T$, so T is negation incomplete.*

Our new theorem is stronger in one respect, weaker in another. But the gain is much more than the loss.

Our new theorem is stronger, because it tells us much more about the character of the undecidable Gödel sentence – namely it has fairly minimal quantifier complexity. The unprovable sentence G_T is a Π_1 sentence of arithmetic. We'll return to say more about this in the next chapter.

Our new theorem is weaker, however, as it only applies to p.r. axiomatized theories, not to (effectively) axiomatized theories more generally. Indeed as we will see in §12.6, Gödel's own original theorems strictly speaking only apply to p.r. axiomatized theories. But that's no real loss. After all, what would a theory look like that was effectively axiomatized but *not* p.r. axiomatized? It would mean that we could e.g. only tell what's an axiom on the basis of an open-ended search: but that would require an *entirely* unnatural way of specifying the theory's axioms in the first place. As we noted at the end of §10.3, any normally presented effectively axiomatized theory will be p.r. axiomatized.

So while we *can* go on to beef up our new result to make it apply as generally as the announced Theorem 1 – and we will say more about this in the next Interlude – there is really only limited interest in doing so. The real force of the (semantic) First Incompleteness Theorem is already captured by Theorem 37.

12   The First Incompleteness Theorem, syntactic version

We now use the same construction of a Gödel sentence as in the previous chapter to show again that PA is incomplete, but this time making only syntactic assumptions. And then we show how to generalize this syntactic version of the incompleteness theorem.

12.1 ω-completeness, ω-consistency

We need to define two key notions. We'll assume in this section that we are dealing with theories whose language includes the language of basic arithmetic. And take all the quantifiers mentioned to run over the natural numbers.[1]

First, then,

Defn. 49. *A theory T is ω-incomplete iff, for some open wff $\varphi(x)$, T can prove $\varphi(\overline{n})$ for each natural number n, but T can't go on to prove $\forall x \varphi(x)$.*

We saw in §5.6 that Q is ω-incomplete: that's because it can prove each instance of $0 + \overline{n} = \overline{n}$, but can't prove $\forall x(0 + x = x)$. We added induction to Q hoping to repair as much ω-incompleteness as we could: but, as we'll see, PA remains ω-incomplete, assuming it is consistent.[2]

Second, we want the following idea:

Defn. 50. *A theory T is ω-inconsistent iff, for some open wff $\varphi(x)$, T can prove each $\varphi(\overline{n})$ and T can also prove $\neg \forall x \varphi(x)$.*

Or, entirely equivalently of course, we could say that T is ω-inconsistent if, for some open wff $\psi(x)$, $T \vdash \exists x \psi(x)$, yet for each number n we have $T \vdash \neg \psi(\overline{n})$.

Note that ω-inconsistency, like ordinary inconsistency, is a syntactically defined property: it is characterized in terms of what wffs can be proved by the theory, not in terms of what the wffs mean. Note too that, in a classical context,

---

[1] If necessary, therefore, read $\forall x \varphi(x)$ as a restricted quantifier $\forall x(Nx \rightarrow \varphi(x))$, where 'N' picks out the numbers from the domain of the theory's native quantifiers (see Defn. 11).

[2] Why the 'ω' in 'ω-incomplete'? Because 'ω' is a standard label for the set of natural numbers (when equipped with their usual ordering).

ω-consistency – defined of course as not being ω-inconsistent! – trivially implies plain consistency. That's because T's being ω-consistent is a matter of its *not* being able to prove a certain combination of wffs, which entails that T can't prove *all* wffs, hence T can't be inconsistent.

Now compare and contrast. Suppose T can prove $\varphi(\bar{n})$ for each n. T is ω-incomplete if it can't prove something we'd then also like it to prove, namely $\forall x \varphi(x)$. While T is ω-inconsistent if it can actually prove the *negation* of what we'd like it to prove, i.e. it can prove $\neg \forall x \varphi(x)$.

So ω-incompleteness in a theory of arithmetic is a regrettable weakness. But ω-inconsistency is a Very Bad Thing (not as bad as outright inconsistency, maybe, but still bad enough). For evidently, a theory T that can prove each of $\varphi(\bar{n})$ and yet also proves $\neg \forall x \varphi(x)$ is just not going to be an acceptable candidate for regimenting arithmetic.

Bring semantic ideas back into play for a moment. Suppose T's standard numerals denote the numbers and the quantifier here runs over all and only the natural numbers. Then it can't be the case that each of $\varphi(\bar{n})$ is true and yet $\neg \forall x \varphi(x)$ is true too. So our ω-inconsistent T can't be sound.

Given that we want formal arithmetics to have theorems which *are* all true on a standard interpretation, we must therefore want ω-consistent arithmetics. And given that we think e.g. PA *is* sound on its standard interpretation, we are committed to thinking that it *is* ω-consistent.

12.2   The First Theorem for PA – the syntactic version

G is by definition the diagonalization of the open wff U $=_{\mathrm{def}}$ $\neg \exists x\, \mathsf{Prfd}(x, y)$, i.e. G is the wff $\neg \exists x\, \mathsf{Prfd}(x, \ulcorner U \urcorner)$ (see §11.1). And now recall Defn. 46: the wff Prfd by hypothesis doesn't just express *Prfd* but *captures* it. Using this fact about Prfd, we can again show that PA does not prove G, but this time *without* making the semantic assumption that PA is sound:

Theorem 39. *If* PA *is consistent,* PA $\nvdash$ G.

Proof. Suppose that PA $\vdash$ G, i.e. (i) PA $\vdash$ $\neg \exists x\, \mathsf{Prfd}(x, \ulcorner U \urcorner)$.

If G has a proof, then there is some super g.n. m that codes its proof. But G is the diagonalization of the wff with g.n. $\ulcorner U \urcorner$. Hence, by definition, $Prfd(m, \ulcorner U \urcorner)$. Since Prfd captures the relation *Prfd*, it follows that PA $\vdash$ $\mathsf{Prfd}(\bar{m}, \ulcorner U \urcorner)$. So, existentially quantifying, we have (ii) $\exists x\, \mathsf{Prfd}(x, \ulcorner U \urcorner)$.

Hence, combining (i) and (ii), the supposition that PA $\vdash$ G entails that PA is inconsistent. Therefore, if PA is consistent, PA $\nvdash$ G.   ⊠

We'll now show that PA also can't prove the *negation* of G, again without assuming PA's soundness:

Theorem 40. *If* PA *is ω-consistent,* PA $\nvdash$ $\neg$G.

Proof. Suppose PA is ω-consistent and PA $\vdash \neg$G. We derive a contradiction, and the theorem follows.

Given our supposition that PA is ω-consistent, it is consistent. Hence, given our supposition that PA proves $\neg$G, it can't prove G. So no number m is the super g.n. of a proof for G. But, again, G is the diagonalization of the wff with g.n. $\ulcorner$U$\urcorner$. Hence, for every number m, $Prfd(m, \ulcorner$U$\urcorner)$ is false.

Since Prfd captures the relation $Prfd$ and $Prfd(m, \ulcorner$U$\urcorner)$ is false for each m, by the definition of capturing we have (i) for each m, PA $\vdash \neg$Prfd$(\overline{m}, \ulcorner$U$\urcorner)$.

But our supposition PA $\vdash \neg$G is equivalent to (ii) PA $\vdash \exists$xPrfd$(x, \ulcorner$U$\urcorner)$.

Combining (i) and (ii), PA is ω-inconsistent, contradicting our initial supposition. $\boxtimes$

Now recall that G is a Π_1 sentence. That observation put together with what we've shown in this section gives us the following portmanteau result:

Theorem 41. *If* PA *is consistent, then there is a* Π_1 *sentence* G *such that* PA $\nvdash$ G, *and if* PA *is* ω-consistent PA $\nvdash \neg$G. *Hence, assuming* ω-consistency *and so consistency,* PA *is negation incomplete.*

12.3   Two corollaries

We pause to note two corollaries. The first is very important:

Theorem 42. *If* PA *is consistent, it is* ω-incomplete.

Proof. Assume PA is consistent. By Theorem 39, it doesn't prove G. So no number m is the super g.n. of a proof of G. Hence, exactly as in the proof of Theorem 40, (i) for each m, PA $\vdash \neg$Prfd$(\overline{m}, \ulcorner$U$\urcorner)$. But the Theorem that G is unprovable in PA is trivially equivalent to (ii) PA $\nvdash \forall$x$\neg$Prfd$(x, \ulcorner$U$\urcorner)$.

The combination of (i) and (ii) shows that PA is ω-incomplete. $\boxtimes$

Let's linger over this result. The incompleteness theorem tells us that there are truths of basic arithmetic which PA can't prove. Reporting it like *that*, however, leaves open the possibility that these will be recondite truths to be found in some remote corner of arithmetic of little interest. But not so. Gödel has found incompleteness surprisingly close to home, in what PA can prove about effectively computable properties of numbers. In particular,

> There's a certain effectively decidable property F which in fact every number has. As we'd hope of a competent formal theory of arithmetic, PA can track what we can compute – so it too can correctly prove of any given number m that m is F. But PA can't take what would seem to be the *very* modest extra step of proving the explicit generalization that *every* number is F.

We have an effectively computable relation *Prfd* (which with only a little effort can be provided with a definitional chain showing it to primitive recursive);

and so there is an effectively computable property F of *not* standing in the *Prfd* relation to the particular number $\ulcorner U \urcorner$. We now know that every number has property F. And indeed (i), for each particular m, PA can correctly prove $\neg\mathsf{Prfd}(\overline{m}, \ulcorner U \urcorner)$. But PA can't take the extra step (ii) and prove the corresponding universal quantification $\forall x \neg\mathsf{Prfd}(x, \ulcorner U \urcorner)$. This sort of incompleteness – as we will see – also affects any system of arithmetic satisfying some natural conditions. Which is surely remarkable.[3]

For future use, here is the other corollary we want to mention:

Theorem 43. *If* PA *is consistent, then* PA $+ \neg$G *(the theory you get by adding* $\neg$G *as an additional axiom) is also consistent but is* ω-*inconsistent.*

Proof. Assume PA is consistent. If PA $+ \neg$G were inconsistent, then PA would prove G, contrary to Theorem 39. So PA $+ \neg$G is also consistent.

Now, since the expanded theory can prove everything PA can prove, we know as before that (i) for each m, PA $+ \neg$G $\vdash \neg\mathsf{Prfd}(\overline{m}, \ulcorner U \urcorner)$.

But just by the definition of $\neg$G, (ii) PA $+ \neg$G $\vdash \exists x\mathsf{Prfd}(x, \ulcorner U \urcorner)$.

And (i) and (ii) together imply that PA $+ \neg$G is ω-inconsistent. ⊠

12.4 Generalizing the proof

The proof for Theorem 41 evidently generalizes. Suppose T is a p.r. axiomatized theory which (perhaps after introducing some new vocabulary by definitions) *contains* Q – in the obvious sense that the language of T includes the language of basic arithmetic, and T can prove every Q-theorem. Then, assuming we are working with a normal scheme for Gödel-numbering wffs of T, the relation $Prfd_T(m, n)$ which holds when m numbers a T-proof of the diagonalization of the wff with number n will be primitive recursive again.

Since T can prove everything Q proves, T will be able to capture the p.r. relation $Prfd_T$ by a Σ_1 wff Prfd_T. Just as we did for PA, we'll be able to construct the corresponding Π_1 wff G_T. And, exactly the same arguments as before will then show, more generally,

Theorem 44. *If* T *is a consistent p.r. axiomatized theory which contains* Q, *then there will be a* Π_1 *sentence* G_T *such that* $T \nvdash \mathsf{G}_T$, *and if* T *is* ω-*consistent,* $T \nvdash \neg\mathsf{G}_T$. *Hence, assuming* ω-*consistency and so consistency,* T *is negation incomplete.*

T will also be ω-incomplete in the same way as PA is. And note, this gives us another incompletability theorem: if we keep chucking more and more additional axioms at our theory T, it will still remain negation incomplete, unless it stops being ω-consistent or stops being p.r. axiomatized.

---

[3] In his classic book on *Gödel's Incompleteness Theorems* (OUP 1992, p. 73), Raymond Smullyan remarks that, while the average mathematician who isn't a logician has heard about theories of arithmetic not being able to decide everything, relatively few have heard of this remarkable ω-incompleteness.

When people refer to the *First Incompleteness Theorem* (without qualification), they typically mean something like our last Theorem, deriving incompleteness from *syntactic* assumptions. Let's re-emphasize that last point. Being p.r. axiomatized is a syntactic property; containing Q is a matter of Q's axioms again being adopted as axioms or being derivable, a syntactic property; being consistent here is a matter of no contradictory pair φ, $\neg\varphi$ being derivable; being ω-consistent is another syntactic property as we stressed before. The chains of argument that lead to our Theorem depend just on the given syntactic assumptions, via e.g. the proof that Q can capture all p.r. functions – another claim about a syntactically definable property. That is why we are calling this the *syntactic* incompleteness theorem.

Of course, we are *interested* in these various syntactically definable properties because of their semantic relevance: for example, we care about the idea of capturing p.r. functions because we are interested in what an interpreted theory might be able to prove in the sense of establish-as-true. But it is one thing for us to have a semantic motivation for being interested in a certain concept, it is another thing for that concept to have semantic content. Again, we are guided to the construction of a Gödel sentence G_T by considerations of what it as-it-were says: but its interpretation isn't involved at all in the proof of its formal undecidability from syntactic assumptions.

12.5 Comparing old and new syntactic incompleteness theorems

Compare Theorem 44 with our initially announced

Theorem 2. *Suppose T is a formal axiomatized theory whose language contains the language of basic arithmetic. Then, if T is consistent and can prove a certain modest amount of arithmetic (and has a certain additional property that any sensible formalized arithmetic will share), there will be a sentence G_T of basic arithmetic such that $T \nvdash G_T$ and $T \nvdash \neg G_T$, so T is negation incomplete.*

Our new theorem fills out the old one in various respects. It fixes the 'modest amount of arithmetic' that T is assumed to contain and it also spells out the 'additional desirable property' of ω-consistency which we previously left mysterious. Further it tells us more about the undecidable Gödel sentence – namely it has minimal quantifier complexity, i.e. it is a Π_1 sentence of arithmetic. Our new theorem is weaker, however, as it only applies to p.r. axiomatized theories, not to effectively axiomatized theories more generally. But as we've already noted at the end of the last chapter, that that's not much loss. (And again, if we insist, we can in fact go on to make up the shortfall: see the following Interlude.)

12.6 Gödel's own Theorem

As we said, Theorem 44, or something like it, is what people usually mean when they speak without qualification of 'The First Incompleteness Theorem'.

But since the stated theorem refers to Robinson Arithmetic Q (developed by Robinson in 1950), and Gödel didn't originally know about that (in 1931), our version can't be quite what Gödel originally proved. But it is a near miss. Let's explain.

Looking again at our analysis of the syntactic argument for incompleteness, we see that we are interested in theories which extend Q *because we are interested in theories which can capture p.r. relations like Prfd*. It's being able to capture *Prfd* that is the crucial condition for our proof to go through. So let's say

Defn. 51. *A theory T is* p.r. adequate *if it can capture all primitive recursive functions and relations.*

Then, instead of mentioning Q, let's instead explicitly write in the requirement of p.r. adequacy. So, by just the same arguments,

Theorem 45. *If T is a p.r. adequate, p.r. axiomatized theory whose language includes L_A, then there is a Π_1 sentence φ such that, if T is consistent then $T \nvdash \varphi$, and if T is ω-consistent then $T \nvdash \neg\varphi$.*

And *this* is pretty much Gödel's own general version of the incompleteness result, and has as much historical right as any to be called *Gödel's* First Theorem.

("Hold on! If *that's* the original First Theorem, we don't need to do the hard work showing that Q and PA are p.r. adequate, do we?" Well, yes and no. No, proving *this* original version of the Theorem of course doesn't depend on proving that any particular theory is p.r. adequate. But yes, showing that this Theorem has real bite, showing that it does actually apply to arithmetics like PA, does depend on proving that these arithmetics are indeed p.r. adequate.)

Thus, in his 1931 paper, Gödel proves that the formal system P – which is his simplified version of the hierarchical type-theory of *Principia Mathematica* – has a formally undecidable Π_1 sentence.[4] This comes from his Theorem VI, which he immediately generalizes:

> In the proof of Theorem VI no properties of the system P were used besides the following:
>
> 1. The class of axioms and the rules of inference (that is, the relation 'immediate consequence') are [primitive] recursively definable (as soon as we replace the primitive signs in some way by the natural numbers).
>
> 2. Every [primitive] recursive relation is definable [i.e., in our terms, is 'capturable'] in the system P.

---

[4]Or as Gödel put it, the undecidable sentence is 'of Goldbach type'. The allusion here is to Goldbach's conjecture that every even number other than two is the sum of two primes. The claim that the particular number n is an even number other than two and is the sum of two primes is expressible by a Δ_0 wff (why?). So Goldbach's conjecture, the universal quantification of this claim about n, is expressible by a Π_1 wff.

Therefore, in every formal system that satisfies the assumptions 1 and 2 and is ω-consistent, there are undecidable propositions of the form $[\forall x F(x)]$, where F is a [primitive] recursively defined property of natural numbers, and likewise in every extension of such a system by a recursively definable ω-consistent class of axioms.

Which gives us a version of our Theorem 45.

Interlude

Time to take stock again. In the last Interlude, we signposted the route forward which would take us to the point where we could prove the our two principal versions of the First Theorem. We have now established the version with a semantic assumption:

Theorem 37. *If T is a sound p.r. axiomatized theory whose language contains the language of basic arithmetic, then there will be a true Π_1 sentence G_T such that $T \nvdash G_T$ and $T \nvdash \neg G_T$, so T is negation incomplete.*

And here is our version requiring instead a syntactic assumption:

Theorem 44. *If T is a consistent p.r. axiomatized theory which contains Q, then there will be a Π_1 sentence G_T such that $T \nvdash G_T$, and if T is ω-consistent, $T \nvdash \neg G_T$. Hence, assuming ω-consistency and so consistency, T is negation incomplete.*

We won't review again the proofs of these main theorems; it will be more useful in this Interlude to pause to review a couple of issues we have left dangling.

We have already promised to say something about the first issue: how to beef up these stated Theorems (which talk of p.r. axiomatized theories) in order to establish versions of our originally announced Theorems 1 and 2 (which talked of effectively axiomatized theories more generally).

A second issue may very well have occurred to you. We can raise it like this: "OK, PA isn't complete. And we have seen that the incompleteness is not in an exotically remote region of arithmetic, but in what PA can prove about computable properties. But still, our undecidable Gödel sentence results from a perhaps not-very-natural coding construction. Are there examples of more *natural* arithmetic propositions, propositions we might already be mathematically interested in for their own sake, which are also independent of PA?"

Let's take these issues in turn. We can't here give full discussions: we just aim to touch very briefly on some key ideas.

(a)   The primitive recursive functions are the numerical functions that can be computed without open-ended searches. So to get a more inclusive class of computable functions, the obvious move is to allow such searches. The standard way of implementing this move is by introducing a *minimization* operator.

Take a simple example. Suppose $g(x, y)$ is a computable function – and suppose too that g is *regular*, meaning that for all x there is at least one y such that $g(x, y) = 0$. Then we can define a function f as follows: $f(x)$ is the least value of y such that $g(x, y) = 0$. So defined, f is computable by a 'do until' open-ended search. To determine the value of $f(x)$ for a particular input x, start computing $g(x, y)$ for increasing values of y (starting from zero); do this until you get to the first value of y such that $g(x, y) = 0$, and output the value of y. Our assumption that g is regular ensures that the search terminates for any given input x. Which makes $f(x)$ a total computable function. Working from this example, we can arrive at the general idea of defining a computable function by *regular minimization*.

Then we say that a function is (total) *recursive* if it is a function that can be defined from our stock of initial functions by applications of composition, primitive recursion, and/or regular minimization. The recursive functions are again effectively computable; and the primitive recursive functions are a subclass of them. As you would expect, we will also say a property or relation is recursive if and only if its characteristic function is recursive.

(b)   Next, note that we can render the minimization operator using the logical apparatus of L_A (we will just need the existential quantifier, identity and the less-than-or-equals relation, which are all available). Since we can already express composition and primitive recursion in L_A, it is therefore easy to strengthen the proof that (a language including) L_A can express every primitive recursive function to show that it can express every recursive function. Similarly, we can strengthen the proof that (a theory including) Q can capture every primitive recursive function to show that it can capture every recursive function.

Now, a *p.r. axiomatized* theory T is one for which e.g. the property of being an axiom is primitive recursive, and consequently the relation Prf_T is primitive recursive. Likewise, a theory T is *recursively axiomatized* is one for which e.g. the property of being an axiom is recursive, and consequently the relation Prf_T is recursive. Given that L_A can express and Q can capture every recursive function, it is straightforward to check that our proofs of Theorems 37 and 44 will go through just as before if we now replace 'primitive recursive' with plain 'recursive', giving us a pair of strengthened Theorems.

All this is pretty plain sailing. But where does it take us? True, these strengthened Theorems apply a bit more widely. But we still haven't quite got back to Theorems 1 and 2, which talk of effectively axiomatized theories, not of recursively axiomatized theories.

(c)   However, we can make the final connection if we accept what is known as *Church's Thesis*, which boldly claims that the effectively computable numerical functions are in fact *exactly* the recursive functions. If we accept Church's Thesis, then we will recover Theorems 1 and 2.

Is Church's Thesis the sort of thing that itself can be *proved* correct, given the imprecision of the informal concept of effective computability? That is debatable. But certainly, every serious attempt to pin down a sharply defined class of effectively computable functions ends up locating the same class of recursive functions. For example, Alan Turing famously aimed to analyse the notion of computation and arrived at the definition of computability by a Turing machine: but we can quite easily prove that every numerical (total) function computable by a Turing machine is recursive, and vice versa. Because of results like this, Church's Thesis commands almost universal assent. For our purposes, we will assume it is true. Which gives us Theorems 1 and 2.

(d)   We know that there are computable functions which aren't primitive recursive from our diagonal argument for Theorem 24. But can we give a natural example of a computable, recursive-but-not-primitive-recursive function?

Here's one construction. Consider the following two-place p.r. functions: f_1, i.e. *sum* (repeated applications of the successor function); f_2, i.e. *product* (repeated sums); f_3, i.e. *exponentiation* (repeated products). As their respective arguments grow, the value of f_1 of course grows comparatively slowly, f_2 grows faster, f_3 faster still.

This sequence of p.r. functions can obviously be continued. Next comes f_4, the *super-exponential*, defined by repeated exponentiation:

$$x \Uparrow 0 = x$$
$$x \Uparrow Sy = x^{x \Uparrow y}.$$

Thus, for example, $3 \Uparrow 4$ is $3^{3^{3^{3^3}}}$ with a 'tower' of four exponents. Similarly, we can define f_5 (super-duper-exponentiation, i.e. repeated super-exponentiation), f_6 (repeated super-duper-exponentiation), and so on. The values of these p.r. functions grow faster and faster as their arguments are increased.

So now let's introduce what we'll call (slightly inaccurately) the Ackermann function $A(x) = f_x(x, x)$. The value of $A(x)$ grows *explosively*, running away ever faster as x increases. Now, we can show that for any given p.r. function $g(x)$ there is an n such that its rate of growth as x increases is no faster than that of $f_n(x, x)$. But $f_x(x, x)$ eventually grows faster than any particular $f_n(x, x)$, when $x > n$. Hence the function A can't be primitive recursive. Yet $A(x)$ is evidently effectively computable – proceed to the definition of f_x, and compute its value for the inputs x, x. And with just a bit of work A can be directly shown to be recursive.

(e)   Fast-growing functions are an old topic of mathematical interest, ever since G. H. Hardy wrote a classic paper about them in 1904. A fast-growing function like our function A was first introduced by Wilhelm Ackermann in 1928 as an example of a computable, but not primitive recursive, function. A is relatively tame, though, as fast-growing functions go. Like any recursive function, it can be expressed by an L_A wff $\mathsf{A}(\mathsf{x}, \mathsf{y})$; and – a further feature – PA 'knows' that A is a total function, giving a (unique) output for any input, i.e. PA can *prove* $\forall\mathsf{x}\exists!\mathsf{y}\mathsf{A}(\mathsf{x}, \mathsf{y})$.

However, we can define wilder and wilder fast-growing functions, to the point where they lose that property of being provably-total-according-to-PA. For example, we can define the so-called Goodstein function G; this is a super-fast-growing recursive function which – like any recursive function – can be canonically expressed in PA by an L_A wff $G(x, y)$. But this time, PA doesn't 'know' that G is a total function, meaning that PA *can't* prove $\forall x \exists! y G(x, y)$.[5]

It would take us too far afield to define the Goodstein function and explain the proof that it really is a total function and the proof that it isn't provably-total-according-to-PA. But without going into those details, we now have a gesture towards the answer to the question we raised at the beginning of this Interlude. Are there arithmetical truths of ordinary mathematical interest which are expressible in PA but undecidable by the theory? Yes, there is a whole family of such truths, truths about super-fast-growing functions.

(f)   Here, though, we must leave these fascinating issues.[6] So where now? We will spend four more chapters still in the neighbourhood of the First Theorem, arriving at a couple of other important theorems, Rosser's Theorem and Tarski's Theorem. We will also be able to make a link back to the lovely Theorem 7, that showed that a consistent sufficiently strong theory is undecidable.

But you could, on a first reading, jump forward to the final three chapters, where we at last discuss the Second Theorem.

---

[5]Note that, unlike the Gödel sentence, *this* new unprovable truth is not Π_1 but has a greater quantifier complexity. It is a nice question whether there are any 'natural' Π_1 truths which are unprovable in PA.

[6]For a lot more on the topics touched on in this Interlude, see *IGT2*. For example, Ch. 30 discusses Goodstein's Theorem and the Goodstein function. Ch. 38 discusses recursive functions in general and the Ackermann function in particular. Turing and Turing machines feature in Chs. 41 and 42, while Chs. 44 and 45 explore Church's Thesis.

13 The Diagonalization Lemma

In this chapter, we establish the so-called Diagonalization Lemma, and use it to re-prove the syntactic version of the First Theorem. In Chapter 15 we use the Lemma again to prove Rosser's Theorem (a technical improvement on the syntactic First Theorem which allows us to drop the assumption of ω-consistency); and in Chapter 16 the Lemma is put to work to prove Tarski's Theorem (a much deeper result, about the 'undefinability of truth').

The Lemma is important, then; and – as you will see – its derivation is not difficult to follow once you grasp the key construction. However, it is perhaps worth pausing at the end of the chapter to describe a nice idea due to Kripke which gets us a tweaked version of the Diagonalization Lemma more or less for free, and we show how this version can also be used to prove incompleteness.

A general remark. To make a smooth fit with our earlier discussions, we will continue to focus on p.r. axiomatized theories. However, the Interlude has given us an inkling of why the same results will carry over to effectively axiomatized theories more generally. But we won't need to make use of this point, except in Chapter 14.

13.1 Two quick reminders

Before getting down to business, it's useful to recall the definition of capturing a function:

Defn. 37. *The theory T captures the one-place function f by the open wff $\psi(\mathsf{x}, \mathsf{y})$ iff, for any m, n,*

  i. if $f(m) = n$, then $T \vdash \psi(\overline{m}, \overline{n})$,
  ii. if $f(m) \neq n$, then $T \vdash \neg\psi(\overline{m}, \overline{n})$.
  iii. $T \vdash \exists!\mathsf{y}\psi(\overline{m}, \mathsf{y})$.

And this time, let's note explicitly that (i) and (iii) together imply

  iv. if $f(m) = n$, then $T \vdash \forall\mathsf{x}(\psi(\overline{m}, \mathsf{x}) \leftrightarrow \mathsf{x} = \overline{n})$.

Next, let's restate Theorem 32 about the *diag* function, and now apply Theorems 27 and 28 to add an important clause about expressing and capturing it (we assume as usual that we have a normal Gödel-numbering scheme in place for the relevant theory T):

Theorem 46. *If T is a p.r. axiomatized theory which contains* Q, *there is a p.r. function* $diag_T(n)$ *which, when applied to a number n which is the g.n. of some T-wff with one free variable, yields the g.n. of that wff's diagonalization, and yields 0 otherwise. And, as with any p.r. function, T can express and capture this function by a Σ_1 wff* $\mathsf{Diag}_T(\mathsf{x}, \mathsf{y})$.

13.2 The Diagonalization Lemma

(a) Let's think again how we constructed our Gödel sentence G_T.

There were two key steps. First, we tweaked the relation *Prf*, and defined *Prfd*, where *Prfd*(m, n) holds iff *Prf*$(m, diag(n))$.

Now, *Prf* will be canonically expressed and captured by a wff Prf. So how can we construct Prfd from Prf? If only we had a function expression $\mathsf{diag}(\mathsf{x})$ to render the function *diag* – then we could express *Prfd* by $\mathsf{Prf}(\mathsf{x}, \mathsf{diag}(\mathsf{y}))$. But we are in fact using a relational expression abbreviated $\mathsf{Diag}(\mathsf{x}, \mathsf{y})$ to render *diag*. To get the same effect, we'll need to fix on a canonical definition like this:

$$\mathsf{Prfd}(\mathsf{x}, \mathsf{y}) =_{\text{def}} \exists \mathsf{z}(\mathsf{Prf}(\mathsf{x}, \mathsf{z}) \wedge \mathsf{Diag}(\mathsf{y}, \mathsf{z})).$$

Then second – after applying quantification and negation – we arrived at a one-place predicate which we then diagonalized to get our Gödel sentence.

(b) So: we have found a wff G_T which is true if and only if it is *unprovable in T*. We now show that this is an instance of a much more general phenomenon. Roughly: take *any* arithmetizable property F; then we can use diagonalization to construct a T-sentence which is true if and only it has property F (and T can prove that equivalence).

More carefully, we have the two-part *Diagonalization Lemma* (Rudolf Carnap noted a version of (i): he is often, but I think wrongly, attributed a version of (ii) as well). As usual, assume a normal Gödel-numbering scheme is in play: then,

Theorem 47. *If T is a p.r. axiomatized theory which contains* Q, *and φ is a one-place open sentence of T's language, then there is a sentence δ such that (i) $\delta \leftrightarrow \varphi(\ulcorner \delta \urcorner)$ is true, and moreover (ii) $T \vdash \delta \leftrightarrow \varphi(\ulcorner \delta \urcorner)$.*

We use a similar two-step construction. Dropping the subscript 'T', first we put

$$\alpha =_{\text{def}} \exists \mathsf{z}(\mathsf{Diag}(\mathsf{x}, \mathsf{z}) \wedge \varphi(\mathsf{z})).$$

Then we let δ be the diagonalization of α. So,

$$\delta =_{\text{def}} \exists \mathsf{z}(\mathsf{Diag}(\ulcorner \alpha \urcorner, \mathsf{z}) \wedge \varphi(\mathsf{z})).$$

We now show that (i) and (ii) are true for this δ. We just apply definitions and infer easy consequences.

Proof for (i). Because diagonalizing α yields δ, by definition $diag(\ulcorner \alpha \urcorner) = \ulcorner \delta \urcorner$. Since Diag expresses *diag*, $\mathsf{Diag}(\ulcorner \alpha \urcorner, \ulcorner \delta \urcorner)$ will be true; indeed, $\mathsf{Diag}(\ulcorner \alpha \urcorner, \mathsf{z})$ is *only* satisfied by $\ulcorner \delta \urcorner$.

This implies that $\exists z(\mathsf{Diag}(\ulcorner\alpha\urcorner, z) \wedge \varphi(z))$ is true if and only if $\ulcorner\delta\urcorner$ satisfies $\varphi(z)$. In other words, δ is true if and only if $\varphi(\ulcorner\delta\urcorner)$ is true. $\boxtimes$

Proof for (ii). Since Diag captures *diag* in T, (iv) from Defn. 37 above tells us that *if $diag(\ulcorner\alpha\urcorner) = \ulcorner\delta\urcorner$, then $T \vdash \forall z(\mathsf{Diag}(\ulcorner\alpha\urcorner, z) \leftrightarrow z = \ulcorner\delta\urcorner)$.*

But, as we noted before, $diag(\ulcorner\alpha\urcorner) = \ulcorner\delta\urcorner$. Hence we can conclude that, indeed, $T \vdash \forall z(\mathsf{Diag}(\ulcorner\alpha\urcorner, z) \leftrightarrow z = \ulcorner\delta\urcorner)$.

So T proves the equivalence of $\mathsf{Diag}(\ulcorner\alpha\urcorner, z)$ and $z = \ulcorner\delta\urcorner$. Therefore T can also prove the equivalence of δ, i.e. $\exists z(\mathsf{Diag}(\ulcorner\alpha\urcorner, z) \wedge \varphi(z))$, with $\exists z(z = \ulcorner\delta\urcorner \wedge \varphi(z))$. But the latter is in turn trivially equivalent to $\varphi(\ulcorner\delta\urcorner)$.

Hence $T \vdash \delta \leftrightarrow \varphi(\ulcorner\delta\urcorner)$. $\boxtimes$

A bit of jargon: by a mild abuse of mathematical terminology,

Defn. 52. *If δ is such that $T \vdash \delta \leftrightarrow \varphi(\ulcorner\delta\urcorner)$, then it is standardly said to be a* fixed point *for φ.*

So the Diagonalization Lemma – or rather, part (ii) of it – is often called the Fixed Point Theorem: for appropriate theories T, every one-place open sentence has a fixed point.

13.3   Incompleteness from the Diagonalization Lemma

Suppose as usual that $Prf_T(m, n)$ is the relation which holds just if m numbers a T proof of a sentence with g.n. n (we continue to assume that we have a normal Gödel-numbering scheme in place). If T is p.r. axiomatized and contains Q, this relation is p.r. decidable and can be expressed and captured in T by a wff $\mathsf{Prf}_T(\mathsf{x}, \mathsf{y})$. And now we pick up again an idea we first met in §3.6:

Defn. 16. *Put $\mathsf{Prov}_T(\mathsf{x}) =_{\mathrm{def}} \exists z\mathsf{Prf}_T(z, \mathsf{x})$ (where the quantifier runs over all the numbers in the domain). Then $\mathsf{Prov}_T(\bar{n})$, i.e. $\exists z\mathsf{Prf}_T(z, \bar{n})$, is true iff some number Gödel-numbers a T-proof of the wff with Gödel-number n, i.e. is true just if the wff with code number n is a T-theorem. So $\mathsf{Prov}_T(\mathsf{x})$ is naturally called a* provability predicate.

$\mathsf{Prov}_T(\ulcorner\varphi\urcorner)$ is true, therefore, just when φ is a theorem.

And now we have a general result about fixed points for the *negation* of such a provability predicate:

Theorem 48. *Suppose T is p.r. axiomatized, contains Q, and some sentence γ is a fixed point for $\neg\mathsf{Prov}_T$; in other words, suppose $T \vdash \gamma \leftrightarrow \neg\mathsf{Prov}_T(\ulcorner\gamma\urcorner)$. Then (i) if T is consistent, $T \nvdash \gamma$. And (ii) if T is ω-consistent, $T \nvdash \neg\gamma$.*

Again, to avoid unsightly rashes of subscripts, let's drop subscript 'T's. Then we can argue like this:

Proof for (i). Suppose $T \vdash \gamma$. Then, since $T \vdash \gamma \leftrightarrow \neg\mathsf{Prov}_T(\ulcorner\gamma\urcorner)$, we have $T \vdash \neg\mathsf{Prov}(\ulcorner\gamma\urcorner)$. But if there *is* a proof of γ, then for some m, $Prf(m, \ulcorner\gamma\urcorner)$, so

$T \vdash \mathsf{Prf}(\overline{m}, \ulcorner\gamma\urcorner)$, since T captures Prf by Prf. Hence $T \vdash \exists x\,\mathsf{Prf}(x, \ulcorner\gamma\urcorner)$, i.e. we also have $T \vdash \mathsf{Prov}(\ulcorner\gamma\urcorner)$, making T inconsistent. So if T is consistent, $T \nvdash \gamma$. ⊠

Proof for (ii). Suppose $T \vdash \neg\gamma$. Then, since $T \vdash \gamma \leftrightarrow \neg\mathsf{Prov}_T(\ulcorner\gamma\urcorner)$, we have $T \vdash \mathsf{Prov}(\ulcorner\gamma\urcorner)$, i.e. $T \vdash \exists x\,\mathsf{Prf}(x, \ulcorner\gamma\urcorner)$. Given T is consistent and proves $\neg\gamma$, there is no proof of γ, i.e. for every m, not-$Prf(m, \ulcorner\gamma\urcorner)$, whence for every m, $T \vdash \neg\mathsf{Prf}(\overline{m}, \ulcorner\gamma\urcorner)$. So we have a $\psi(x)$ such that T proves $\exists x\psi(x)$ while it refutes each instance $\psi(\overline{m})$.

Hence, if T is consistent and $T \vdash \neg\gamma$, T is ω-inconsistent. So if T is ω-consistent (and hence consistent), $T \nvdash \neg\gamma$. ⊠

But part (ii) of the Diagonalization Lemma tells us in particular that

Theorem 49. *If T is p.r. axiomatized, contains* Q, *then there exists a sentence γ such that $T \vdash \gamma \leftrightarrow \neg\mathsf{Prov}_T(\ulcorner\gamma\urcorner)$.*

So putting our two theorems together we get incompleteness again. And note that all that is required so far is that we are working with a wff Prf_T that captures Prf_T. Any fixed point γ for the negation of any provability predicate will give us a formally undecidable wff.[1] Moreover, since γ is undecidable, so is $\neg\mathsf{Prov}_T(\ulcorner\gamma\urcorner)$, and that is a Π_1 wff (because Prov_T is Σ_1).

So putting everything together, we recover again the syntactic First Theorem, Theorem 44.

13.4 Proving our old G_T is a fixed point for $\neg\mathsf{Prov}_T$

How does our new proof of the syntactic incompleteness theorem relate to the old one?

Recall once more, our Gödel sentence G_T is the diagonalization of $\mathsf{U}(y) =_{\mathrm{def}} \neg\exists x\,\mathsf{Prfd}(x, y)$. G_T is true if and only if unprovable-in-T. This fact can now be *expressed* inside T itself, by the wff $G_T \leftrightarrow \neg\mathsf{Prov}_T(\ulcorner G_T\urcorner)$ (and compare our informal Theorem 5, stated but not proved in §3.6). But T doesn't just express this fact but can *prove* it too. We have

Theorem 50. *If T is p.r. axiomatized and contains* Q, $T \vdash G_T \leftrightarrow \neg\mathsf{Prov}_T(\ulcorner G_T\urcorner)$.

In other words, our G_T is one of those fixed points for $\neg\mathsf{Prov}_T$. The proof is elementary, appealing to definitions and simple logical manipulations:

Proof. Dropping subscripts again, $Prfd(m, n)$ holds iff $Prf(m, diag(n))$. As noted in §13.2, we can therefore fix on the following canonical definition:

$$\mathsf{Prfd}(x, y) =_{\mathrm{def}} \exists z(\mathsf{Prf}(x, z) \wedge \mathsf{Diag}(y, z)).$$

---

[1] Warning. Some authors call *any* such fixed point for a predicate $\neg\mathsf{Prov}_T$ built from *any* wff which captures Prf_T a Gödel sentence for T. That's fine, as long as you are alert to the fact that not everything that is true of canonical Gödel sentences in the narrow sense we introduced in the preceding chapters is true of Gödel sentences in this new wider sense.

Diagonalizing U yields G. Hence, by the definition of *diag*, we have $diag(\ulcorner U \urcorner) = \ulcorner G \urcorner$. Since by definition Diag captures *diag*, it follows – by the remark at the beginning of §13.1 – that

(*) $T \vdash \forall z(\mathsf{Diag}(\ulcorner U \urcorner, z) \leftrightarrow z = \ulcorner G \urcorner)$.

We can then do some elementary manipulations inside theory T:

$$
\begin{array}{llll}
\mathsf{G} & \leftrightarrow & \neg \exists x \, \mathsf{Prfd}(x, \ulcorner U \urcorner) & \text{(by defn of G)} \\
& \leftrightarrow & \neg \exists x \exists z (\mathsf{Prf}(x,z) \wedge \mathsf{Diag}(\ulcorner U \urcorner, z)) & \text{(by defn of Prfd)} \\
& \leftrightarrow & \forall z(\mathsf{Diag}(\ulcorner U \urcorner, z) \rightarrow \neg \exists x \, \mathsf{Prf}(x,z)) & \text{(logic)} \\
& \leftrightarrow & \forall z(\mathsf{Diag}(\ulcorner U \urcorner, z) \rightarrow \neg \mathsf{Prov}(z)) & \text{(by defn of Prov)} \\
& \leftrightarrow & \forall z(z = \ulcorner G \urcorner \rightarrow \neg \mathsf{Prov}(z)) & \text{(using (*))} \\
& \leftrightarrow & \neg \mathsf{Prov}(\ulcorner G \urcorner) & \text{(logic)} \quad\boxtimes
\end{array}
$$

13.5 An easier Diagonalization Lemma?

An afterword for enthusiasts. Our proof of the standard form of the Diagonalization Lemma was not particularly difficult. But it was a bit messy and relied on a not-immediately-obvious initial construction. So it is interesting to note that we can get the Lemma or a close variant more or less for free in a couple of different ways.

One way is to use a devious form of Gödel-numbering, one especially cooked up so that wffs with one free variable get a special run of code numbers, directly chosen to make the Diagonalization Lemma come out true. Now, this initially smacks of cheating – for aren't we hiding away a key idea in the deviant numbering scheme? But there is more to be said (though not here).

A second way is simpler, and claimed by Saul Kripke to be more natural. Though this time, the application of his variant Lemma takes rather more care and attention.[2]

Suppose, then, that T is the usual kind of theory, p.r. axiomatized and containing Q (and so incorporating the language of basic arithmetic). We can effectively list its open wffs with the one free variable x in the usual way $\varphi_0(x), \varphi_1(x), \varphi_2(x), \varphi_3(x), \ldots$. Now add to T's language a corresponding infinite sequence of *new* constants $c_0, c_1, c_2, c_3, \ldots$, and expand our assumed Gödel-numbering scheme for T to give code numbers to each of the new constants.[3] Then:

Defn. 53. *The* Kripke extension T^K *of a theory T whose open wffs with free variable* x *are* $\varphi_i(x)$ *is the result of adding all the new constants* c_i *to T, together with accompanying new axioms* $c_i = \ulcorner \varphi_i(c_i) \urcorner$.

---

[2] Kripke's construction in this section has long been folklore but was only published in 2020, in his 'Gödel's Theorem and direct self-reference' (*arXiv* 2010.11979).

[3] Fine print: this harmlessly assumes that our scheme is like the one introduced in §10.1 in having an infinite sequence of as-yet-unused code numbers available to code the new constants. We can always arrange this to be so, if only by the low trick of doubling all the Gödel numbers in our initially assumed scheme, leaving us with all the odd numbers to code the c_i!

The idea, then, is to introduce constants c_i governed by axioms which ensure that a wff $\varphi_i(c_i)$ is – via the coding – more directly 'about' itself, without the usual diagonalization trickery.

We need to read into the record six quick and easy observations:

O1. T^K is still p.r. axiomatized. We can still effectively determine (without open-ended searches) whether a given wff is an axiom – why?

O2. Any wff without c-constants which is provable in T^K is already provable in T. If we add new constants which are in effect just shorthand for certain standard numerals (different constants, different numerals), then the only new wffs we can prove will be wffs with those constants.

O3. Any wff of T^K with some c-constant(s) is provably-in-T^K equivalent to a wff without c-constants. Just use the c-axioms and Leibniz's Law to replace a constant with a standard numeral.

O4. For any wff α of T's language, if $T^K \nvdash \alpha$, then $T \nvdash \alpha$. Trivially, if the theory with extra axioms can't prove α, then the original theory can't.

O5. If T is consistent, so is T^K. Again, adding distinct new constants for distinct numerals can't make a consistent theory inconsistent.

O6. If T is ω-consistent, so is T^K. We prove the contrapositive. Suppose T^K is ω-inconsistent, so there is an open wff $\psi(x)$, such that $T^K \vdash \exists x \psi(x)$ yet for each number n we have $T^K \vdash \neg\psi(\bar{n})$. Apply (O3) to ψ if necessary to get a provably equivalent ψ' without any of the new constants, and then $T^K \vdash \exists x \psi'(x)$ while for each number n we have $T^K \vdash \neg\psi'(\bar{n})$. But since these wffs in ψ' are free of c-constants, we can apply (O2) to get $T \vdash \exists x \psi'(x)$ while for each number n we have $T \vdash \neg\psi'(\bar{n})$, which makes the original T ω-inconsistent.

We now have the following variant of (the second part) of the original Diagonal Lemma:

Theorem 51. *If T is a p.r. axiomatized theory which contains Q, and φ is a one-place open sentence of T's language, then there is a sentence δ of T's Kripke extension T^K such that $T^K \vdash \delta \leftrightarrow \varphi(\ulcorner\delta\urcorner)$.*

Proof. Trivially, φ will be some φ_i. By the relevant c-axiom, $T^K \vdash \varphi_i(c_i) \leftrightarrow \varphi_i(\ulcorner\varphi_i(c_i)\urcorner)$. So just put $\delta = \varphi_i(c_i)$ and we are done. ⊠

That was easy! Now take the case where $\varphi(x)$ is $\neg\mathsf{Prov}_{T^K}(x)$, the negation of the canonical provability predicate for T^K. This is formed in the (constant-free) language of basic arithmetic, i.e. is a one-place predicate already available in T. So we can invoke our theorem to show that there is in particular a fixed point γ of the language of T^K such that $T^K \vdash \gamma \leftrightarrow \neg\mathsf{Prov}_{T^K}(\ulcorner\gamma\urcorner)$.

And now we can apply Theorem 48. Since we have a fixed point γ for $\neg\mathsf{Prov}_{T^K}$, we know that (i) if T^K is consistent, $T^K \nvdash \gamma$. And (ii) if T^K is ω-consistent, $T^K \nvdash \neg\gamma$.

Fine. But this doesn't quite get us back to the desired incompleteness theorem for our original theory T. For a start, the fixed point γ (on our current construction) involves one of those new constants, so γ won't belong to the language of T. So we now need to use our observations above.

By (O3), we know that there is a γ' without new constants which is provably equivalent to γ, so (i') if T^K is consistent, $T^K \nvdash \gamma'$; and (ii') if T^K is ω-consistent, $T^K \nvdash \neg\gamma'$. So then applying (O5) and (O4), we get (i'') if T is consistent, so is T^K, so $T^K \nvdash \gamma'$, and hence $T \nvdash \gamma'$. And applying (O6) and (O4) we get (ii'') if T is ω-consistent so is T^K, so $T^K \nvdash \neg\gamma'$, hence $T \nvdash \neg\gamma'$. Which gives us the First Incompleteness Theorem for T again.

In summary, then. To get incompleteness from a Diagonalization Lemma, whether our way or Kripke's, you rely on the key Theorem 48. With our original Lemma, you get to apply Theorem 48 'neat'. If you use Kripke's construction to arrive at an easier variant Lemma (adding constants stipulated to do the necessary work), then this time you have to take a bit of care to massage away the constants after you've applied Theorem 48. So you pays your money and you takes your choice!

14 Undecidability

In this short chapter, we pause to make use of Theorem 50, and we link back to our early Theorem 7 that showed that a consistent, sufficiently strong, effectively axiomatized theory is undecidable. We also discuss the so-called *Entscheidungs-problem* (i.e. 'decision problem' for first-order logic).

14.1 Undecidability again

Here again is

Theorem 50. *If T is p.r. axiomatized and contains* Q, $T \vdash \mathsf{G}_T \leftrightarrow \neg\mathsf{Prov}_T(\ulcorner\mathsf{G}_T\urcorner)$.

But we now know from the last Interlude that this sort of result can be strength-ened. Since Q can capture all the recursive properties and functions (not just the *primitive* recursive properties and functions), we will also have a strengthened theorem

Theorem 52. *If T is recursively axiomatized and contains* Q,
$$T \vdash \mathsf{G}_T \leftrightarrow \neg\mathsf{Prov}_T(\ulcorner\mathsf{G}_T\urcorner).$$

We should now note a simple corollary of this last theorem. Suppose Prov_T not only expresses but *captures* the property of being a T-theorem. By the definition of capturing, if φ is a non-theorem, we would then have $T \vdash \neg\mathsf{Prov}_T(\ulcorner\varphi\urcorner)$. Hence, in particular, since G_T is a non-theorem, $T \vdash \neg\mathsf{Prov}_T(\ulcorner\mathsf{G}_T\urcorner)$. But, given Theorem 52, that would imply G_T is a theorem, which it isn't so long as T is consistent. Hence:

Theorem 53. *If T is consistent, recursively axiomatized and contains* Q, *then* Prov_T *does* not *capture the property of being a T-theorem.*

It follows that

Theorem 54. *A consistent, recursively axiomatized theory T which contains* Q *is not recursively decidable.*

That is to say, the property of being (the code number for) a T-theorem is not recursive. For it were, it could be canonically captured in the usual way by our provability predicate Prov_T.

Now, that doesn't quite give us Theorem 7. But we can make the connection if we also accept Church's Thesis – as we should! For that, remember, asserts that the effectively decidable properties are just the recursive ones. And accepting this, the last theorem becomes: a consistent effectively axiomatized theory which contains Q is not effectively decidable. And we know that containing Q is about the weakest way of being 'sufficiently strong'. Which takes us back, more or less, to the earlier result that a consistent, sufficiently strong theory, effectively axiomatized theory is undecidable.

14.2 The *Entscheidungsproblem*

Q is such a *very* simple theory we might quite reasonably have hoped that there *would* be some mechanical way of telling which wffs are and which aren't its theorems. But we now know that there isn't; and, in a sense, you can blame the underlying first-order logic. For, assuming Q's consistency, we get *Church's Theorem* as an immediate corollary of Q's recursive undecidability:

Theorem 55. *The property of being a theorem of first-order logic is not recursively decidable.*

Gödel number the wffs of (your favourite version of) first-order logic; there is no recursive function which, given a code number, returns a 0/1, yes/no, verdict on the wff's theoremhood.

Proof. Since Q is has only a finite number of axioms, we can wrap them together into a single conjunction, $\hat{Q}$. And then, trivially, $Q \vdash \varphi$ if and only if $\vdash \hat{Q} \to \varphi$; i.e. we can prove φ inside Q if and only if a certain related conditional is logically provable from no assumptions. So if (1) the relevant first-order logic built into Q were recursively decidable, then (2) we could decide whether the conditional $\hat{Q} \to \varphi$ is a logical theorem, hence (3) we could effectively decide whether φ is a Q-theorem. But Theorem 54 rules that out. So (1) must be false – first order logic must be undecidable. ⊠

Hilbert and Ackermann's *Grundzüge der theoretischen Logik* (originally published in 1928) is the first recognizably modern logic textbook, still well worth reading. They posed the *Entscheidungsproblem*, the problem of coming up with an effective method for deciding of an arbitrary sentence of first-order logic whether it is valid or not. Theorem 55 tells us that there is no *recursive* function that takes (the g.n. of) a sentence and returns a 0/1, yes/no, verdict on theoremhood. This doesn't *quite* answer Hilbert and Ackerman's problem as posed. But Church's Thesis can again be invoked to bridge the gap between our last theorem and this:

Theorem 56. *The property of being a theorem of first-order logic is indeed not effectively decidable.*

15 Rosser's Theorem

In another comfortingly short chapter, we now redeploy the Diagonalization Lemma to prove Rosser's Theorem. This tweaks the syntactic First Theorem so that we no longer need the assumption of ω-consistency. This is a nice technical result you ought to know about, but is fiddly to prove.

15.1 Tweaking the provability predicate, Rosser-style

One half of the syntactic First Theorem requires the assumption that we are dealing with a theory T which is not only consistent but also ω-consistent. But we can now improve on this. Following J. Barkley Rosser (in 1936), we can construct a Rosser sentence R_T such that we only need to assume T is plain consistent in order to show that R_T is formally undecidable.

Start by considering the relation $\overline{Prf}_T(m, n)$ which holds when m numbers a T-proof of the *negation* of the wff with number n. This relation is obviously p.r. given that Prf_T is. Hence, just as Prf_T can be expressed and captured by a wff $\mathsf{Prf}_T(\mathsf{x}, \mathsf{y})$, assuming T has the usual properties, $\overline{Prf}$ can be expressed and captured by a wff $\overline{\mathsf{Prf}}_T(\mathsf{x}, \mathsf{y})$.

Take then the following expression:

Defn. 54. $\mathsf{RPrf}_T(\mathsf{x}, \mathsf{y}) =_{\mathrm{def}} \mathsf{Prf}_T(\mathsf{x}, \mathsf{y}) \wedge (\forall \mathsf{w} \leq \mathsf{x}) \neg \overline{\mathsf{Prf}}_T(\mathsf{w}, \mathsf{y})$.

A pair of numbers m, n will satisfy this complex predicate if m numbers the proof of a wff with number n, and no smaller number numbers the proof of that wff's negation.

It is worth remarking that this complex predicate RPrf_T still captures the relation Prf_T, assuming T is consistent. Why?

1. Suppose, supressing subscripts, $Prf(m, n)$. Then (i) $T \vdash \mathsf{Prf}(\overline{\mathsf{m}}, \overline{\mathsf{n}})$.

   Moreover, given that T is consistent, since the wff with number n has a proof, no number numbers a proof of the negation of that wff. Which means that $\overline{Prf}(0, n)$, $\overline{Prf}(1, n)$, $\overline{Prf}(3, n)$, ... $\overline{Prf}(m, n)$ are all *false*.

   Hence, because $\overline{\mathsf{Prf}}$ captures $\overline{Prf}$, we have each of $T \vdash \neg \mathsf{Prf}(0, \overline{\mathsf{n}})$, $T \vdash \neg \mathsf{Prf}(1, \overline{\mathsf{n}})$, $T \vdash \neg \mathsf{Prf}(2, \overline{\mathsf{n}})$, ..., $T \vdash \neg \mathsf{Prf}(\overline{\mathsf{m}}, \overline{\mathsf{n}})$.

   Therefore, we will also have (ii) $T \vdash (\forall \mathsf{v} \leq \overline{\mathsf{m}}) \neg \mathsf{Prf}(\overline{\mathsf{m}}, \overline{\mathsf{n}})$ – since Q and hence T 'knows' about bounded quantifiers (see §7.1).

   So from (i) and (ii) we have $T \vdash \mathsf{RPrf}(\overline{\mathsf{m}}, \overline{\mathsf{n}})$.

2. Suppose not-$Prf(m, n)$. Then $T \vdash \neg\mathsf{Prf}(\overline{m}, \overline{n})$, so $T \vdash \neg\mathsf{RPrf}(\overline{m}, \overline{n})$.

And now we can define *the Rosser provability predicate* as follows:

Defn. 55. $\mathsf{RProv}_T(\mathsf{x}) =_{\mathrm{def}} \exists\mathsf{v}\mathsf{RPrf}(\mathsf{v}, \mathsf{x})$.

Then a sentence is Rosser-provable in T – its g.n. satisfies the Rosser provability predicate – if it has a proof (in the ordinary sense) and there's no smaller-numbered proof of its negation.

15.2   Rosser's Theorem

Now we apply the Diagonalization Lemma, not to the negation of a regular provability predicate (which is what we just did to get Gödel's First Theorem again), but to the negation of the Rosser provability predicate. The Lemma then tells us,

Theorem 57. *Given that T is p.r. axiomatized and contains* Q, *then there is a sentence* R_T *such that* $T \vdash \mathsf{R}_T \leftrightarrow \neg\mathsf{RProv}_T(\ulcorner\mathsf{R}_T\urcorner)$.

We call such a sentence R_T a Rosser sentence for T.
    We can now show this key result:

Theorem 58. *Suppose T is a consistent p.r. axiomatized theory which contains* Q *and let γ be any fixed point for* $\neg\mathsf{RProv}_T(\mathsf{x})$. *Then (i) $T \nvdash \gamma$ and (ii) $T \nvdash \neg\gamma$.*

Proof of (i). Just follow the proof of (i) for Theorem 48 – simply change $\mathsf{Prf}/\mathsf{Prov}$ to $\mathsf{RPrf}/\mathsf{RProv}$, and use the fact that RPrf captures *Prf*.   ⊠

Outline proof of (ii). Suppose that $T \vdash \neg\gamma$. By the fixed point biconditional, $T \vdash \mathsf{RProv}(\ulcorner\gamma\urcorner)$.
    Because $\neg\gamma$ is a theorem, for some m, $\overline{Prf}(m, \ulcorner\gamma\urcorner)$, so (a) $T \vdash \overline{\mathsf{Prf}}(\overline{m}, \ulcorner\gamma\urcorner)$.
    Also, since T is consistent, γ is unprovable, so for all n, not-$Prf(n, \ulcorner\gamma\urcorner)$. Hence, in particular, for each n up to m, $T \vdash \neg\mathsf{Prf}(\overline{n}, \ulcorner\gamma\urcorner)$. From which we get $T \vdash (\forall\mathsf{v} \leq \overline{m})\neg\mathsf{Prf}(\mathsf{v}, \ulcorner\gamma\urcorner)$ since T knows about bounded quantifiers. And a bit more work will then give us (b) $T \vdash \forall\mathsf{v}(\mathsf{Prf}(\mathsf{v}, \ulcorner\gamma\urcorner) \to \overline{m} \leq \mathsf{v})$.
    Combining (a) and (b), it immediately follows that $T \vdash \forall\mathsf{v}(\mathsf{Prf}(\mathsf{v}, \ulcorner\gamma\urcorner) \to (\overline{m} \leq \mathsf{v} \wedge \overline{\mathsf{Prf}}(\overline{m}, \ulcorner\gamma\urcorner)))$. And given our definition, that quickly gives us $T \vdash \neg\mathsf{RProv}(\ulcorner\gamma\urcorner)$.
    But now we have shown we can derive both $\mathsf{RProv}(\ulcorner\gamma\urcorner)$ and $\neg\mathsf{RProv}(\ulcorner\gamma\urcorner)$ in T, contradicting its consistency.   ⊠

Putting our last two theorems together, we get the Rosser's Theorem:

Theorem 59. *If T is a consistent, p.r. axiomatized theory which contains* Q, *then there is sentence* R_T *such that, if T is consistent then $T \nvdash \mathsf{R}_T$ and $T \nvdash \neg\mathsf{R}_T$.*

With yet more work, we can beef up the proof idea to show that in fact (as with Gödel's original proof) we can find a Π_1 sentence which is undecidable so long as T is consistent. But we won't do that here![1]

---

[1]Masochists can check out *IGT2*, §25.4

16 Tarski's Theorem

Recall, our Diagonalization Lemma has two parts:

Theorem 47. *If T is a p.r. axiomatized theory which contains Q, and φ is a one-place open sentence of T's language, then there is a sentence δ such that (i) $\delta \leftrightarrow \varphi(\ulcorner\overline{\delta}\urcorner)$ is true, and moreover (ii) $T \vdash \delta \leftrightarrow \varphi(\ulcorner\overline{\delta}\urcorner)$.*

This chapter applies each part of the Lemma, to give a pair of results about truth that are usually packaged together as *Tarski's Theorem*. We arrive at the deep contrast between the notion of truth and the notion of provability which Gödel saw as underlying the incompleteness phenomenon.

16.1 Truth predicates and truth definitions

Recall a familiar thought: 'snow is white' is true iff snow *is* white. Likewise for all other sensible replacements for 'snow is white'. In sum, we can endorse every sensible instance of *'φ' is true iff φ*.[1] And that's because of the very meaning of the informal truth predicate 'true'.

How can we add a truth predicate to an interpreted formal language L which contains the language of basic arithmetic (as in Defn. 11)? Such a language will in general not have quotation marks or the like available; however, we can arithmetize syntax and use code numbers to refer to wffs. Assume that we have fixed on some normal scheme for Gödel numbering L-wffs. Then we can define a corresponding numerical property *True* as follows:

> *True*(n) is true iff n is the g.n. of a true sentence of L.

Now suppose, just suppose, we introduce some expression $\mathsf{T}(\mathsf{x})$ with one free variable which is so defined as to *express* this numerical property *True*. And – allowing for the possibility that we've had to extend L in introducing such an expression – let L^* be the result of adding a new wff $\mathsf{T}(\mathsf{x})$ to our initial language L if necessary. (So for the moment, we leave it open whether L^* is just L, which it would be if a suitable $\mathsf{T}(\mathsf{x})$ is already definable from L's resources.)

Then, by the definition of *True* and of T, we have the following for any sentence φ of the original language L:

---

[1] Forgive the casualness with the quotation marks. Nothing will depend on this!

φ is true iff $True(\ulcorner\varphi\urcorner)$ iff $\mathsf{T}(\ulcorner\varphi\urcorner)$ is true.

Hence, for any L-sentence φ, every corresponding 'T-biconditional'

$$\mathsf{T}(\ulcorner\varphi\urcorner) \leftrightarrow \varphi$$

is true. Which motivates our first main definition:

Defn. 56. *An open L^*-wff $\mathsf{T}(\mathsf{x})$ (where L^* includes L) is a truth predicate for L iff for every L-sentence φ, $\mathsf{T}(\ulcorner\varphi\urcorner) \leftrightarrow \varphi$ is true.*

And here's a companion definition:

Defn. 57. *A theory Θ (with language L^* which includes L) is a formal truth theory for L iff it provides an L^*-wff $\mathsf{T}(\mathsf{x})$ such that $\Theta \vdash \mathsf{T}(\ulcorner\varphi\urcorner) \leftrightarrow \varphi$ for every L-sentence φ.*

Equally often, a truth theory for L is called a 'definition of truth for L'.

In sum, a truth predicate T for L is a predicate of a perhaps extended language that applies to (the Gödel numbers for) exactly the true L-sentences, and so *expresses* truth. And a truth theory for L is a theory built in a perhaps extended language that *proves* all the T-biconditionals for L sentences.

So far, that's just a sequence of (natural enough) definitions. Now for our first big result.

16.2 The undefinability of truth

Suppose T is an arithmetical theory with language L. The question arises: could T be competent to define truth *for its own language* (i.e., can T already encompass a truth theory for L)? And an answer is immediate – call this Tarski's Theorem on the undefinability of truth:

Theorem 60. *No consistent p.r. axiomatized theory T which contains Q can define truth for its own language.*

Proof. Assume T defines truth for L using an open L-sentence $\mathsf{T}(\mathsf{x})$. Since T has the right properties, part (ii) of the Diagonalization Lemma applies. Therefore we can apply the Lemma in particular to $\neg\mathsf{T}(\mathsf{x})$, so there must be some sentence L such that

   1. $T \vdash \mathsf{L} \leftrightarrow \neg\mathsf{T}(\ulcorner\mathsf{L}\urcorner)$.

According to T then, L is a Liar sentence, which (as it were) says that it is false! But, by our initial assumption that T is a truth theory for L using T as its formal truth predicate, we also have

   2. $T \vdash \mathsf{T}(\ulcorner\mathsf{L}\urcorner) \leftrightarrow \mathsf{L}$.

But (1) and (2) together entail that T is inconsistent, contrary to hypothesis. So our assumption must be wrong: T can't define truth for its own language. $\boxtimes$

16.3 Curry's Paradox and Tarski's Theorem

Our proof that a theory T (satisfying the usual conditions) can't define truth for its own language was essentially this: if T could define truth, it would get entangled with the Liar Paradox. It is fun and instructive to give another proof of the same theorem: this time, we show that, if T could define truth for its own language, it would get entangled with Curry's Paradox. But what's that?

We seemingly can construct a sentence that says e.g. 'if this sentence is true, then the moon is made of green cheese' (after all, haven't I just constructed one?). Let's symbolize the whole sentence by δ, and let's put ψ for 'the moon is made of green cheese'.

So by construction, δ supposedly holds exactly when, if δ, then ψ. And now we can argue as follows – just using the familiar inference rules for the conditional and biconditional:

1.	$\delta \leftrightarrow (\delta \to \psi)$	By definition of δ (?)
2.	δ	Supposition
3.	$\delta \to \psi$	From 1, 2
4.	ψ	From 2, 3
5.	$\delta \to \psi$	Conditional proof from 2 to 4
6.	δ	From 1, 5
7.	ψ	From 5, 6

So we have proved that the moon is made of green cheese!

This sort of argument was already known to medieval logicians. But it was rediscovered by Haskell Curry in 1942, and these days – in one form or another – is known as *Curry's Paradox*. Why 'paradox'? Because, at first sight, truths of the kind (1) seem to be available to us, assuming we can construct self-referential sentences at all (and these are surely often harmless, as in 'This sentence contains five words'). Yet evidently something has gone wrong.

Is there anything problematic *after* the starting assumption at line (1)? Seemingly not: for we have only used intuitively secure inference moves. We have simply derived one-way conditionals from a biconditional, and then reasoned about conditionals using modus ponens and conditional proof. So, the blame surely has to fall on the original assumption that there is such a δ as makes (1) hold.

But what is wrong with this assumption? Let's dig deeper. We took the sentence 'if I am true, then the moon is made of green cheese', symbolized the whole conditional as δ, symbolized the consequent of the conditional as ψ, and then asserted that we have

1. $\delta \leftrightarrow (\delta \to \psi)$.

However, that elided two steps. Strictly speaking, if δ symbolises the whole conditional, the antecedent of the conditional should be $True(\delta)$, and our initial equivalence will then be

E. $\delta \leftrightarrow (True(\delta) \to \psi)$

And then we need to appeal to the equivalence

T. $True(\delta) \leftrightarrow \delta$

to arrive at the troublesome (1). So which of (E) and (T) should we give up?

We are not going to tackle this question in general. But let's consider the argument in a formal context. So now assume that both

  I. T is a p.r. axiomatized theory which contains Q;
  II. T can define truth for its own language using the truth predicate $\mathsf{T}(\mathsf{x})$.

Then we can argue as follows.

Take a sentence ψ of T, and consider the open sentence $\mathsf{T}(\mathsf{x}) \to \psi$. Given (I), we have the Diagonalization Lemma, so there will indeed be some δ such that

  1. $T \vdash \delta \leftrightarrow (\mathsf{T}(\overline{\ulcorner\delta\urcorner}) \to \psi)$

And then since, by (II), T can define truth,

  2. $T \vdash \mathsf{T}(\overline{\ulcorner\delta\urcorner}) \leftrightarrow \delta$
  3. $T \vdash \delta \leftrightarrow (\delta \to \psi)$ From 1, 2, using propositional logic in T
  4. $T \vdash \psi$ Arguing in T from 3 as for Curry's Paradox

But ψ was arbitrarily chosen. So T can prove any sentence and hence is inconsistent. So if T is consistent and our assumption (I) about T is true, then (II) is false. Which is just Tarski's Theorem 60 again.

16.4   The inexpressibility of truth

Our first Tarskian theorem puts limits on what a nice theory can *prove* about truth. But we can go further: there are limits on what a theory's language can even *express* about truth.

Consider our old friend L_A for the moment, and suppose that there is an L_A truth predicate T_A that expresses the corresponding truth property $True_A$. Then part (i) of the Diagonalization Lemma, applies to $\neg\mathsf{T}_A(\mathsf{x})$ (for the proof of this semantic part depended only on the fact that we were dealing with a language which contains the language of basic arithmetic). So in particular, there will be some L_A sentence L such that

  1. $\mathsf{L} \leftrightarrow \neg\mathsf{T}_A(\overline{\ulcorner\mathsf{L}\urcorner})$.

is true. But, by the assumption that T_A is a truth predicate for L_A,

  2. $\mathsf{T}_A(\overline{\ulcorner\mathsf{L}\urcorner}) \leftrightarrow \mathsf{L}$

must be true too. (1) and (2) immediately lead to contradiction again. Therefore our supposition that T_A expresses the property of being a true L_A sentence has to be rejected.

The argument generalizes. Take any language L which includes the language of basic arithmetic, so that the first part of the Diagonalization Lemma is provable. Call that an arithmetically adequate language. Then by the same argument we get Tarski's Theorem limiting the expressibility of truth:

Theorem 61. *No predicate of an arithmetically adequate language L can express the numerical property* True_L *(i.e. the property of numbering a truth of L).*

16.5 The Master Argument for incompleteness?

Our second Tarskian result tells us that while you can express *syntactic* properties of a sufficiently rich formal theory of arithmetic (like provability) inside a theory via Gödel numbering, you can't express some key *semantic* properties (like truth) inside the same theory. And this points to a particularly illuminating take on the argument for incompleteness.

For example: PA-provability *is* expressible in L_A (by a provability predicate Prov), but truth-in-L_A *isn't* expressible in L_A. Hence truth in L_A isn't provability in PA. So assuming that PA is sound so that everything provable in it is true, this means that there must be truths of L_A which PA can't prove. Similarly, of course, for other nice theories.

And in a way, we might well take this to be *the* Master Argument for incompleteness, revealing the true roots of the phenomenon. Gödel himself wrote (in response to a query)

> I think the theorem of mine that von Neumann refers to is ... that a complete epistemological description of a language A cannot be given in the same language A, because the concept of truth of sentences in A cannot be defined in A. *It is this theorem which is the true reason for the existence of undecidable propositions in the formal systems containing arithmetic.* I did not, however, formulate it explicitly in my paper of 1931 but only in my Princeton lectures of 1934. The same theorem was proved by Tarski in his paper on the concept of truth.[2]

In sum, as we emphasized before, arithmetical truth and provability in this or that formal system must peel apart.

"How does that statement of Gödel's square with the importance that he placed on the syntactic version of the First Theorem?" Well, Gödel himself was a realist about mathematics (certainly he was in his later philosophical writings; and he claimed that this had been his view all along). So, the story goes, *he* believed in the real existence of mathematical entities, and believed that our theories (at least aim to) deliver truths about them. But that wasn't the

---

[2]Gödel's letter is quoted in Solomon Feferman's very interesting 'Kurt Gödel: conviction and caution', reprinted in Feferman's *In the Light of Logic* (OUP 1998).

dominant belief among those around him concerned with foundational matters. As I put it before, for various reasons the very idea of truth-in-mathematics was under some suspicion at the time. So even if semantic notions were at the root of Gödel's insight, it was extremely important for him – given the intended audience – to show that you don't need to deploy those semantic notions to prove incompleteness.

"But does this mean that, if *we* are happy enough with the use of semantic premisses, we needn't worry about tackling the complications of proving the syntactic version of the First Theorem?" No. For a start, as we are about to see, it is reflection on the syntactic First Theorem which takes us to the Second Theorem ...

17 The Second Theorem and Hilbert's Programme

Time at long last for the Second Theorem. The basic idea is straightforward, the devil is in the details of a full proof (more about that in the next chapter).

Now, we didn't have to do much work to explain why the First Theorem is so interesting – how odd and unexpected to find that even the truths of basic arithmetic resist being completely pinned in a nicely axiomatized theory! But we do need to do rather more scene-setting to help bring out the significance of the Second Theorem. So in this chapter we also give a cartoon sketch of some history.

17.1 Defining Con_T

When talking about various axiomatized theories T in earlier chapters, we didn't specify a particular formulation of the deductive system built into T (other than that the resulting theory counts as p.r. axiomatized). The logic may or may not have a built-in *absurdity constant* like the conventional '$\bot$'. So henceforth, let's use the absurdity sign in the following way:

Defn. 58. *'$\bot$' is T's built-in absurdity constant if it has one, or else it is an abbreviation for '$0 = \bar{1}$'.*

Assuming T contains Q, T of course proves $0 \neq \bar{1}$. So on either reading of '$\bot$', if T proves $\bot$, it is inconsistent. And conversely, if T has a standard classical (or intuitionistic) logic and T is inconsistent, then on either reading it will prove $\bot$.

Assuming again that T contains the language of basic arithmetic, and we have a normal Gödel-numbering scheme in place, then (as we now know) we can express provability-in-T in a canonical way by an open T-wff Prov_T, such that $\mathsf{Prov}_T(\ulcorner\varphi\urcorner)$ is true exactly when φ is a T-theorem.

Hence the wff $\neg\mathsf{Prov}_T(\ulcorner\bot\urcorner)$ is true if and only if T *doesn't* prove $\bot$, hence (given what we've just said) is true if and only if T is consistent. That evidently motivates the definition

Defn. 59. Con_T *abbreviates* $\neg\mathsf{Prov}_T(\ulcorner\bot\urcorner)$.

For obvious reasons, the arithmetic sentence Con_T is called a (canonical) *consistency sentence* for T.[1] Since Prov_T is Σ_1, Con_T is Π_1.

Let's have an example for future use. Take your favourite nicely axiomatized set theory – as it might be, Zermelo-Fraenkel theory plus the Axiom of Choice, the standard theory ZFC. We can define the language of basic arithmetic in the language of set theory. So ZFC can itself express provability-in-ZFC using a wff of the theory, $\mathsf{Prov}_{\mathsf{ZFC}}$, and from that we can form the consistency sentence $\mathsf{Con}_{\mathsf{ZFC}}$ which is true iff ZFC is consistent; moreover, this sentence will be an arithmetic sentence (meaning a ZFC version of a sentence of basic arithmetic).

17.2 The Formalized First Theorem

Back, for a moment, to the First Incompleteness Theorem. Assume as usual that we are dealing with a theory T which is p.r. axiomatized and contains Q. Then one half of the syntactic version of the First Theorem tells us that

(1) If T is consistent, then G_T is not provable in T.

And given what we said in the previous section, we have a natural way of expressing (1) *inside the formal theory T itself*, i.e. by the conditional

(2) $\mathsf{Con}_T \to \neg\mathsf{Prov}_T(\ulcorner \mathsf{G}_T \urcorner)$.

Call this wff the *Formalized First Theorem*.

Now let's reflect that, once we have made the cunning move of constructing G_T, the informal reasoning for the First Theorem is in fact *very* elementary. We certainly needed no higher mathematics at all, just relatively straightforward reasoning about arithmetical codings. So we might well expect that if T can reason about arithmetic sufficiently well, *it should itself be able to reflect that elementary reasoning*, and hence itself *prove* our formalized version of the statement of the First Theorem .

In short, we will hope to have the following key result:

Theorem 62. *If T is p.r. axiomatized and contains enough arithmetic, then* $T \vdash \mathsf{Con}_T \to \neg\mathsf{Prov}_T(\ulcorner \mathsf{G}_T \urcorner)$.

Of course, that's a skeleton version: we'll want to flesh out what counts as 'enough arithmetic'; but we'll do that in the next chapter. But here's an initial thought. Maybe T will need to be stronger that Q, because it will presumably need to be able to use induction in generalizing about arithmetized syntax; but containing PA ought to be enough.

---

[1] There are alternatives for defining consistency sentences. Suppose the wff $\mathsf{Contr}(x, y)$ captures the p.r. relation which holds between two numbers when they code for a contradictory pair of sentences, i.e. one codes for some sentence φ and the other for $\neg\varphi$. Then we could define Con_T^* to be short for the sentence $\neg\exists x\exists y(\mathsf{Prov}_T(x) \wedge \mathsf{Prov}_T(y) \wedge \mathsf{Contr}(x, y))$, which says that we can't find two T-provable wffs which contradict each other. But, on modest assumptions, Con_T^* is provably equivalent to Con_T. Another equivalent option would be to use as a consistency sentence the formalization of 'there is a sentence which T doesn't prove'. But we will stick to our standard definition.

17.3   From the Formalized First Theorem to the Second Theorem

Assume we have established Theorem 62 (we'll discuss its proof in the next chapter). Then, if T is p.r. axiomatized, and contains enough arithmetic,

1. $T \vdash \mathsf{Con}_T \to \neg\mathsf{Prov}_T(\ulcorner \mathsf{G_T} \urcorner)$

But we know from Theorem 50 that, for a p.r. axiomatized T which contains as little arithmetic as Q,

2. $T \vdash \mathsf{G}_T \leftrightarrow \neg\mathsf{Prov}_T(\ulcorner \mathsf{G_T} \urcorner)$.

Hence from (1) and (2) we have

3. $T \vdash \mathsf{Con}_T \to \mathsf{G}_T$.

We can therefore immediately infer that

4. If $T \vdash \mathsf{Con}_T$, then $T \vdash \mathsf{G}_T$.

But we know from the First Theorem (syntactic version) that

5. If T is consistent, $T \nvdash \mathsf{G}_T$.

So, making the conditions on T explicit again, from (1) we get a (somewhat unspecific) version of Gödel's *Second Incompleteness Theorem*.

Theorem 63. *Suppose T is p.r. axiomatized and contains enough arithmetic: then, if T is consistent, $T \nvdash \mathsf{Con}_T$.*

17.4   The impact of the Second Theorem?

If we could have derived Con_T in the theory T, that would have been no special evidence *for* T's consistency. After all, we can derive anything in an inconsistent theory; so in an inconsistent T we could derive Con_T in particular.

On the other hand, T's failure to prove Con_T is no evidence *against* T's consistency. We have just found another true Π_1 sentence that T cannot prove, to set alongside G_T.

Hence – you might initially suppose – the non-derivability of a canonical statement of T's consistency inside T itself does not show us a great deal.

But that's too fast. For consider this obvious corollary of the Second Theorem:

Theorem 64. *Suppose S is a consistent theory, strong enough for the Second Theorem to apply to it, and W is a weaker fragment of S, then $W \nvdash \mathsf{Con}_S$.*

That's because, if S is strong enough for the Second Theorem to apply to it, it can't prove Con_S. So then, a fortiori, *part* of S can't prove Con_S either.

So, for example, we *can't* take some strong theory like ZFC as the theory S and show that it is consistent by (i) using arithmetic coding for talking about its proofs and then (ii) using uncontentious reasoning already available in some weaker, perhaps purely arithmetical, theory W. Assuming it is consistent, the stronger theory ZFC can't prove Con_{ZFC}. So assuming the weaker arithmetic theory is equivalent to a fragment of set theory, W can't prove Con_{ZFC} either. And *this* is an important result. Why? We need to fill in some historical background.

17.5 Formalization, finitary reasoning, and Hilbert

Think yourself back to the situation in mathematics a bit over a century ago. Classical analysis – the theory of differentiation and integration – has, supposedly, been put on firm foundations. We have, for example, done away with obscure talk about infinitesimals; and we have traded in an intuitive grasp of the continuum of real numbers for the idea of reals defined as 'Dedekind cuts' on the rationals or 'Cauchy sequences' of rationals. The key idea we've used in our constructions is the idea of a *set* of numbers. And we've been very free and easy with that, allowing ourselves to talk of arbitrary sets of numbers, even when there is no statable rule for collecting the numbers into the set.

This freedom to allow ourselves to talk of arbitrarily constructed sets is just one aspect of the increasing freedom that mathematicians have allowed themselves over the second half of the nineteenth century. We have loosed ourselves from the assumption that mathematics should be tied to the description of nature: as Morris Kline puts it, "after about 1850, the view that mathematics can introduce and deal with arbitrary concepts and theories that do not have any immediate physical interpretation ... gained acceptance". For example, Cantor could write in 1883 "Mathematics is in its development entirely free and is only bound in the self-evident respect that its concepts must both be consistent with each other and with [other established concepts]".[2]

It is rather bad news, then, if all this play with freely created concepts, and in particular the fundamental notion of arbitrary sets, actually gets us embroiled in contradiction – as seems to be the case once the set-theoretic paradoxes (like Russell's Paradox) are discovered. What to do?

We might – rather artificially – distinguish two lines of responses to the paradoxes that threaten Cantor's paradise where mathematicians can play freely, which we might suggestively call the *foundationalist* and the *more careful mathematics* responses.

Foundationalist responses to paradox　Consider first the option of seeking external 'foundations' for mathematics.

We could, perhaps, seek to 're-ground' mathematics by confining ourselves again to applicable mathematics which has, as we would anachronistically put it, a model in the natural world so *must* be consistent.

The trouble with this idea is we're none too clear what this re-grounding in the world would involve – for remember, we are thinking back at the beginning of the twentieth century, as relativity and quantum mechanics are emerging, and

---

[2]Morris Kline, *Mathematical Thought from Ancient to Modern Times* (OUP, 1972) p. 1031. Georg Cantor, *Grundlagen einer allgemeinen Mannigfaltigkeitslehre* (Tuebner 1883) §8.

any Newtonian confidence that we had about the real structure of the natural world is being shaken.

So maybe we have to put the option of anchoring mathematics in the physical world aside. But perhaps we could try to ensure that our mathematical constructions are grounded in the mental world, in mental constructions that we can perform and have a secure epistemic access to. This idea leads us to Brouwer's intuitionism. But it depends on an obscure notion of mental construction, and in any case – in its most worked out form – this approach *cripples* a lot of classical mathematics that we thought was unproblematically in good order, rather than giving it a foundation.

What other foundational line might we take? The notion of set that we have used (and seemingly abused) is arguably in some sense a logical notion; perhaps we have got into trouble by going beyond its logical core. How about trying to nail down some incontrovertible logical principles and then find definitions of mathematical notions in purely logical terms, so constraining mathematics to be what we can reconstruct on a firm *logical* footing.

This *logicist* line which we briefly met back in §1.5 has its independent attractions – which is why Frege embarked on it before Russell's and other paradoxes were discovered. But it is problematic in various ways. Remember, we are pretending to be situated a hundred or so years back, and at least at this point modern logic itself isn't in as good a shape as most of the mathematics we are supposedly going to use it to ground (and what might *count* as incontrovertible logic is still pretty obscure). Moreover, as C. S. Peirce saw, it looks as if we are going to need to appeal to mathematically developed ideas in order to develop logic itself; indeed Peirce himself thought that all formal logic is merely mathematics applied to logic.

Still, perhaps we shouldn't give up yet. The proof is in the pudding: let's see if we can actually do the job and reconstruct mathematics on a logical basis as Frege tried to do (but this time, without getting entangled in paradox), and let's write a *Principia Mathematica* ...!

'Mathematical' responses to paradox   Maybe, however, back at the beginning of the twentieth century, we just shouldn't be seeking to give mathematics a prior 'foundation' after all. Consider: the paradoxes arise within mathematics, and to avoid them (the working mathematican might reasonably think) we just need to do the mathematics more carefully. As Peirce – for example – held, mathematics risks being radically distorted if we seek to make it answerable to some outside considerations. We don't need to look outside mathematics (to the world, to mental constructions, or even to logic) for a justification that will guarantee consistency. Perhaps we just need to improve our mathematical practice, in particular by improving the explicitness of our regimentations of mathematical arguments, to reveal the principles we actually use in 'ordinary' mathematics, and to see where the fatal mis-steps must be occurring when we over-stretch these principles in ways that lead to paradox.

How do we improve explicitness? A first step will be to work towards putting the principles that we actually need in mathematics into something approaching the ideal form of an effectively axiomatized formal theory. This is what Zermelo aimed to do in axiomatizing set theory: to locate the principles actually needed for the seemingly 'safe' mathematical constructions needed in grounding classical analysis and other familiar mathematical practice. And when the job is done it seems that *these* principles don't in fact allow the familiar reasoning leading to Russell's Paradox or other set-theoretic paradoxes. So like the foundationalist/logicist camp, the 'do maths better' camp are also keen on regimenting a mathematical theory T into a nice tidy axiomatized format and sharply defining the rules of the game: but the purpose of the axiomatization is quite significantly different. The aim now is not to expose foundations, but just to put our theory into a form that – hopefully – we can now show is indeed consistent and trouble-free.

For note the crucial insight of Hilbert's, which we can sum up like this:

> The axiomatic formalization of a mathematical theory T (whether about sets, widgets, or other whatnots), gives us *new* objects (beyond the sets, widgets, or other whatnots) that are *themselves* apt topics for new mathematical investigations – namely the T-wffs and T-proofs that make up the theory!
>
> And, crucially, when we go metatheoretical like this and move from thinking about *sets* (as it might be) to thinking about the syntactic properties of *formalized-theories-about-sets*, we can be moving from considering suites of *infinite* objects to considering suites of *finite* formal objects (the wffs, and the finite sequences of wffs that form proofs). This means that we might then hope to bring to bear, at the metatheoretical level, entirely 'safe', merely *finitary*, reasoning about these suites of finite formal objects in order to prove the consistency of (say) set theory.

Of course, it is a debatable point what exactly constitutes such ultra-'safe' finitary reasoning. But still, it certainly looks as if we will – for instance – need much, much, less than full set theory to reason (not about *sets* but) about a formalized *theory* of sets as a suite of finite syntactic objects. So we might, in particular, hope with Hilbert – in the do-mathematics-better camp – to be able to use a safe uncontentious fragment of finitary mathematics to prove that our wildly infinitary set theory (ZFC, perhaps) is at least syntactically consistent.

So, at this point (I hope!) you can begin to see the attractions of what's called *Hilbert's Programme* – i.e. the programme of showing various systems of carefully reconstructed infinitary mathematics are contradiction-free by giving consistency proofs using safe, finitary, reasoning about the systems considered as formal objects.

But now, enter Gödel, wielding his Second Theorem. He shows that a theory which contains enough arithmetic can't prove its own consistency, let alone the consistency of a stronger theory which includes arithmetic. Hence, for example, we can't use a weak fragment of ZFC – a version of finitary arithmetic, for example – to prove ZFC consistent by proving Con$_{ZFC}$: not even full-powered ZFC itself can prove Con$_{ZFC}$. Similarly for other strong theories. Which means that the Second Theorem – at least at first blush – sabotages Hilbert's Programme.[3]

Famously, when Hilbert heard of the result at a conference where Gödel first announced it, he was not well pleased . . . .

---

[3]Fine print. The Second Theorem doesn't rule out all possibility of interesting consistency proofs. For a start, it leaves open the possibility of showing that a theory S is consistent by appeal to a theory S' which is weaker in some respects and stronger in others. Gerhard Gentzen, for example, found such a proof of the consistency of PA. But it would take us too far afield to explore this sort of possibility here (though see *IGT2* §32.4).

18 Proving the Second Incompleteness Theorem

In his original 1931 paper, Gödel states a version of the Second Theorem. But he doesn't spell out a full proof. He simply claims that the reasoning for the First Theorem is so elementary that a strong enough theory must be able to replicate the reasoning and prove the Formalized First Theorem; and then he notes that this implies the Second Theorem.[1]

The hard work of taking a strong enough T and checking that it really does prove the Formalized First Theorem was first done for a particular case by David Hilbert and Paul Bernays in their *Grundlagen der Mathematik* of 1939. The details of their proof are – the story goes – due to Bernays, who had discussed it with Gödel during a transatlantic voyage.

Now, Hilbert and Bernays helpfully isolated what we can call *derivability conditions* on the predicate Prov_T, conditions whose satisfaction is enough for a theory T to prove the Formalized First Theorem. Later, Martin H. Löb gave a rather neater version of these conditions, giving us the so-called *HBL conditions* which we invoke in this chapter.

18.1 Sharpening the Second Theorem

Recall: $\mathsf{Prov}_T(y)$ abbreviates $\exists v\, \mathsf{Prf}_T(v, y)$; and Prf_T is a Σ_1 expression. So arguing about provability inside T will involve establishing some general claims involving Σ_1 expressions. And how do we prove general arithmetical claims? Using induction is the default method.

It therefore looks to be a good bet that we will get $T \vdash \mathsf{Con}_T \to \neg\mathsf{Prov}_T(\ulcorner \mathsf{G}_T \urcorner)$ if T at least has Σ_1-induction – meaning that T's axioms include (the universal closures of) all instances of the first-order Induction Schema where the induction predicate φ is no more complex than Σ_1.

In §7.5 we introduced $\mathsf{I}\Sigma_1$ as the standard label for the result of adding that amount of induction to Q. So it is a plausible conjecture that we can show the following, a fleshed-out version of our skeletal Theorem 62:

---

[1] I am told that Gödel's shorthand notebooks at the time suggest that he in fact hadn't then worked out any detailed proof of the first step.

Theorem 65. *If T is p.r. axiomatized and contains $I\Sigma_1$, then T proves the Formalized First Theorem, i.e.* $T \vdash \mathsf{Con}_T \to \neg\mathsf{Prov}_T(\ulcorner G_T \urcorner)$

And if that is right, by the argument of §17.3 we get an improved version of Theorem 63, a shaper Second Theorem:

Theorem 66. *If T is consistent, p.r. axiomatized and contains $I\Sigma_1$, then* $T \nvdash \mathsf{Con}_T$.

So how can we prove Theorem 65?

18.2 The box notation

To improve readability, we introduce some standard notation. We will henceforth abbreviate $\mathsf{Prov}_T(\ulcorner \varphi \urcorner)$ simply by $\Box_T\varphi$ – you can read that as 'φ is provable (in T)'. This is perhaps a bit naughty, as the new notation hides away the Gödel-numbering and the use of standard numerals for the codes. But the notation is standard and is safe enough if we keep our wits about us.[2]

So in particular, $\neg\mathsf{Prov}_T(\ulcorner G_T \urcorner)$ can be abbreviated $\neg\Box_T G_T$. Thus in our new notation, the Formalized First Theorem is $\mathsf{Con}_T \to \neg\Box_T G_T$. Moreover, Con_T can now alternatively be abbreviated as $\neg\Box_T\bot$.

18.3 Proving the Formalized First Theorem

First a standard definition: we will say (dropping subscripts)

Defn. 60. *The HBL derivability conditions* hold in T *if and only if, for any T-sentences φ, ψ,*

C1. If $T \vdash \varphi$, then $T \vdash \Box\varphi$;

C2. $T \vdash \Box(\varphi \to \psi) \to (\Box\varphi \to \Box\psi)$;

C3. $T \vdash \Box\varphi \to \Box\Box\varphi$.

Here, (C1) tells us that if T can prove φ, then (via the relevant Gödel-numbering which enables T to code up claims about provability-in-T), T as it were 'knows' that φ is provable. (C2) tells us that T 'knows' about modus ponens; in other words, if T can show $(\varphi \to \psi)$ is provable, and can show φ is provable, then T can show that ψ is provable too. And (C3) tells us that if T 'knows' it can prove φ, it can code up that proof so it can prove that φ is provable. We might reasonably hope that these conditions will hold for a strong enough theory T.

And we can indeed show that

---

[2]If you are familiar with modal logic, then you will immediately recognize the conventional symbol for the necessity operator. And the parallels and differences between '"$1 + 1 = 2$" is provable (in T)' and 'It is necessarily true that $1 + 1 = 2$' are highly suggestive. These parallels and differences are the topic of 'provability logic', the subject of a contemporary classic, George Boolos's *The Logic of Provability* (CUP, 1993).

Theorem 67. *If T is p.r. axiomatized and contains $\mathsf{I\Sigma}_1$, then the derivability conditions hold for T.*

However, demonstrating this is a seriously tedious task, and I certainly *don't* propose to hack through the annoying technical details. In these notes we will just take this technical theorem as given.

We can also show, considerably more easily, that

Theorem 68. *If T is p.r. axiomatized, contains Q, and the derivability conditions hold for T, then T proves the Formalized First Theorem.*

Proof. This is just a mildly fun exercise in box-juggling, with the target of showing $T \vdash \mathsf{Con}_T \to \neg \Box_T \mathsf{G}_T$.

First, since T is p.r. axiomatized and contains Q, Theorem 50 holds. So, in our new symbolism and dropping subscripts once more, we have $T \vdash \mathsf{G} \leftrightarrow \neg \Box \mathsf{G}$.

Second, since T contains Q and standard logic, we have

$$T \vdash \neg \varphi \to (\varphi \to \bot).$$

(whichever way we read the absurdity constant – see Defn. 58). Given this and the derivability condition (C1), we get

$$T \vdash \Box(\neg \varphi \to (\varphi \to \bot)).$$

So given the derivability condition (C2) and using modus ponens, it follows that for any φ

A. $T \vdash \Box \neg \varphi \to \Box(\varphi \to \bot).$

We now argue as follows:

1.	$T \vdash \mathsf{G} \to \neg \Box \mathsf{G}$	Half of Theorem 50
2.	$T \vdash \Box(\mathsf{G} \to \neg \Box \mathsf{G})$	From 1, given C1
3.	$T \vdash \Box \mathsf{G} \to \Box \neg \Box \mathsf{G}$	From 2, using C2
4.	$T \vdash \Box \neg \Box \mathsf{G} \to \Box(\Box \mathsf{G} \to \bot)$	Instance of A
5.	$T \vdash \Box \mathsf{G} \to \Box(\Box \mathsf{G} \to \bot)$	From 3 and 4
6.	$T \vdash \Box \mathsf{G} \to (\Box\Box \mathsf{G} \to \Box \bot)$	From 5, using C2 and logic
7.	$T \vdash \Box \mathsf{G} \to \Box\Box \mathsf{G}$	Instance of C3
8.	$T \vdash \Box \mathsf{G} \to \Box \bot$	From 6 and 7
9.	$T \vdash \neg \Box \bot \to \neg \Box \mathsf{G}$	Contraposing
10.	$T \vdash \mathsf{Con} \to \neg \Box \mathsf{G}$	Definition of Con ⊠

Now we put the last two theorems together. If T is p.r. axiomatized and contains $\mathsf{I\Sigma}_1$, the derivability conditions will hold (by Theorem 67) and it contains Q. Hence (by Theorem 68) if T is p.r. axiomatized and contains $\mathsf{I\Sigma}_1$, it proves the Formalized First Theorem. Which gives us our target Theorem 65.

18.4 The equivalence of Con_T with G_T

We'll assume throughout this section that T is a p.r. axiomatized theory such that the derivability conditions hold. And first, we will show that in that case, G_T and Con_T are provably equivalent in T.

We have just shown in the previous Theorem that $T \vdash \mathsf{Con} \rightarrow \neg\Box\mathsf{G}$. We can now prove the converse, $T \vdash \neg\Box\mathsf{G} \rightarrow \mathsf{Con}$. Indeed, we have a more general result:

Theorem 69. *For any sentence* φ, $T \vdash \neg\Box\varphi \rightarrow \mathsf{Con}$.

Proof. We argue as follows:

1.	$T \vdash \bot \rightarrow \varphi$	Logic!
2.	$T \vdash \Box(\bot \rightarrow \varphi)$	From 1, given C1
3.	$T \vdash \Box\bot \rightarrow \Box\varphi$	From 2, given C2
4.	$T \vdash \neg\Box\varphi \rightarrow \neg\Box\bot$	Contraposing
5.	$T \vdash \neg\Box\varphi \rightarrow \mathsf{Con}$	Definition of Con ⊠

This little theorem has a rather remarkable corollary. Since T can't prove Con, the theorem tells us that T doesn't entail $\neg\Box\varphi$ for any φ at all. Hence T doesn't ever 'know' that it can't prove φ, even when it can't! In sum, suppose that T satisfies the now familiar conditions: by (C1), T knows all about what it *can* prove; but we have just shown that it knows nothing about what it *can't* prove.

Now, as a particular instance of our last theorem, we have $T \vdash \neg\Box\mathsf{G} \rightarrow \mathsf{Con}$. So putting that together with Theorem 68, we have $T \vdash \mathsf{Con} \leftrightarrow \neg\Box\mathsf{G}$. And now combine *that* with Theorem 50 which tells us that $T \vdash \mathsf{G} \leftrightarrow \neg\Box\mathsf{G}$, and lo and behold we've shown

Theorem 70. *If T is p.r. axiomatized and contains* $\mathsf{I\Sigma_1}$, *then* $T \vdash \mathsf{Con} \leftrightarrow \mathsf{G}$.

This means that, not only do we have $T \nvdash \mathsf{Con}$, we also have (assuming in addition that T is ω-consistent) $T \nvdash \neg\mathsf{Con}$. In other words, Con is formally undecidable by T.

But Con is *not* a wff which is more or less directly 'about' itself. That observation should scotch any lingering suspicion that undecidable sentences provided by Gödelian arguments are all tainted by potentially paradoxical self-reference.

We next prove

Theorem 71. $T \vdash \mathsf{Con} \leftrightarrow \neg\Box\mathsf{Con}$.

Proof. The direction from right to left is an instance of Theorem 69. For the other direction, note that

1.	$T \vdash \mathsf{Con} \rightarrow \mathsf{G}$	Already proved
2.	$T \vdash \Box(\mathsf{Con} \rightarrow \mathsf{G})$	From 1, given C1
3.	$T \vdash \Box\mathsf{Con} \rightarrow \Box\mathsf{G}$	From 2, given C2
4.	$T \vdash \neg\Box\mathsf{G} \rightarrow \neg\Box\mathsf{Con}$	Contraposing
5.	$T \vdash \mathsf{Con} \rightarrow \neg\Box\mathsf{Con}$	Using Thm. 68 ⊠

So: this shows that Con (like G) is also a fixed point of the negated provability predicate (see again Defn. 52 and §13.4).

As we have noted before (§13.3, fn. 1), some authors refer to any fixed point of the negated provability predicate as a Gödel sentence. Fine. That's one way of using the jargon. But if you adopt the broad usage, you must be careful with your informal commentary. For example, not all Gödel sentences in the broad sense indirectly 'say' *I am unprovable*: Con is a case in point.

18.5 Löb's Theorem

Now that we have the derivability conditions in play, let's state and prove a theorem from 1955, due to Martin Löb:

Theorem 72. *If T is p.r. axiomatized and contains $I\Sigma_1$ (so the derivability conditions hold), then for any T-wff φ, if $T \vdash \Box\varphi \to \varphi$ then $T \vdash \varphi$.*

Proof. More box juggling! Assume that, for a given φ,

1. $T \vdash \Box\varphi \to \varphi$.

And next consider the wff $\mathsf{Prov}(x) \to \varphi$. Given our assumptions about T, the Diagonalization Lemma applies. Hence for some δ, T proves $\delta \leftrightarrow (\mathsf{Prov}(\ulcorner\delta\urcorner) \to \varphi)$. Or, using our box notation,

2. $T \vdash \delta \leftrightarrow (\Box\delta \to \varphi)$

And now we can continue as follows:

3.	$T \vdash \delta \to (\Box\delta \to \varphi)$	From 2
4.	$T \vdash \Box(\delta \to (\Box\delta \to \varphi))$	From 3, by C1
5.	$T \vdash \Box\delta \to \Box(\Box\delta \to \varphi)$	From 4, by C2
6.	$T \vdash \Box\delta \to (\Box\Box\delta \to \Box\varphi)$	From 5, by C2
7.	$T \vdash \Box\delta \to \Box\Box\delta$	By C3
8.	$T \vdash \Box\delta \to \Box\varphi$	From 6 and 7
9.	$T \vdash \Box\delta \to \varphi$	From 1 and 8
10.	$T \vdash \delta$	From 2 and 9
11.	$T \vdash \Box\delta$	From 10, by C1
12.	$T \vdash \varphi$	From 9 and 11. $\boxtimes$

This is perhaps rather a surprise. We might have expected that a theory which has a well-constructed provability-predicate should 'think' that, quite generally, if it can prove φ then indeed φ – i.e. we might have expected that in general $T \vdash \Box\varphi \to \varphi$. But not so. A respectable theory T can only prove this if in fact it can *already* show that φ.

By the way, Löb's Theorem answers a question raised by Leon Henkin. We know from the Diagonalization Lemma applied to the *negated* wff $\mathsf{Prov}(x)$ that there is a sentence G such that $T \vdash G \leftrightarrow \neg\mathsf{Prov}(\ulcorner G\urcorner)$, and that a consistent T can't prove G. Similarly, we can apply the Lemma to the *unnegated* wff $\mathsf{Prov}(x)$

to show that there is a sentence H such that $T \vdash H \leftrightarrow \mathsf{Prov}(\ulcorner H \urcorner)$ – so it is as if H says 'I am provable'. Henkin asked: *is* H provable? Well, we now know that it is. For by hypothesis, $T \vdash \mathsf{Prov}(\ulcorner H \urcorner) \to H$, i.e. $T \vdash \Box H \to H$; so $T \vdash H$ by Löb's Theorem.

Löb's Theorem also immediately entails the Second Theorem again. For assume the conditions for Löb's Theorem apply. Then as a special case we get that if $T \vdash \Box \bot \to \bot$ then $T \vdash \bot$. Therefore, if $T \nvdash \bot$, so T is consistent, then $T \nvdash \Box \bot \to \bot$, hence $T \nvdash \neg \Box \bot$, hence (just by definition of Con), $T \nvdash$ Con.

There's a converse too, i.e. the Second Theorem fairly quickly entails Löb's Theorem.[3] So the two theorems come to much the same. And now note how the proof of Löb's Theorem from the biconditional at line (2) to the conclusion at line (12) is very reminiscent of the argument for Curry's Paradox which we met in §16.3; and we can with a bit of massaging make the arguments run more closely in parallel. So the situation is this. What Gödel saw in proving the first theorem, in the broadest terms, is that when we move from talking about truth to talking about provability, thinking about Liar-style sentences can lead not to paradox but to the First Incompleteness Theorem. Much later, Löb spotted that, similarly, when we move from talking about truth to talking about provability, Curry-paradox style reasoning again leads not to paradox but to the Second Theorem.

---

[3]For the proof idea due to Kripke, see *IGT2*, §34.5.

19   A subtle difference

This final chapter highlights an important but easy-to-miss difference between what is involved in proving the first theorem and what the second theorem requires. But we start with an instructive remark on ...

19.1   Consistent theories that 'prove their own inconsistency'

An ω-*consistent* (and so consistent) p.r. axiomatized theory T which contains a little arithmetic can't prove $\neg\mathsf{Con}_T$, as we've seen. By contrast, a consistent but ω-*inconsistent* T can have $\neg\mathsf{Con}_T$ as a theorem. (Being ω-inconsistent, T won't be sound; that's how it can have a false theorem!)

The proof is actually straightforward, once we note a simple lemma. Suppose S and R are two p.r. axiomatized theories, which share a deductive logic; and suppose every axiom of the simpler theory S is also an axiom of the richer theory R. It is then a trivial logical truth that, if the richer theory R is consistent, then the simpler theory S must be consistent too. Assuming that R contains Q, it will be able to formally prove the arithmetical claim that encodes this fact:

Theorem 73. *Under the given conditions, with theory R extending theory S,* $R \vdash \mathsf{Con}_R \to \mathsf{Con}_S$.

So now take our simpler theory S to be PA. Take the richer theory R to be PA augmented by the extra axiom $\neg\mathsf{G}_{\mathsf{PA}}$. We have met this richer theory briefly before: by Theorem 43, R is consistent but ω-inconsistent. Since R trivially proves $\neg\mathsf{G}_{\mathsf{PA}}$, and PA (and hence R) proves G_{PA} is equivalent to $\mathsf{Con}_{\mathsf{PA}}$, R proves $\neg\mathsf{Con}_{\mathsf{PA}}$. So – using our last theorem and modus tollens – R proves $\neg\mathsf{Con}_R$.

Summing that up,

Theorem 74. *Assuming* PA *is consistent, the theory* PA $+ \neg\mathsf{G}_{\mathsf{PA}}$ *is a consistent theory which 'proves' itself inconsistent (i.e. we can derive in it the negation of its own consistency sentence).*

And since R proves $\neg\mathsf{Con}_R$,

Theorem 75. *There can be a* consistent *theory R such that the theory $R + \mathsf{Con}_R$ is* inconsistent.

What are we to make of these apparent absurdities? Well, giving the language of R, i.e. PA $+ \neg G_{PA}$, its standard arithmetical interpretation, the theory is unsound (since the Gödel sentence G_{PA} is true). So we shouldn't trust what R says about anything, especially about its own inconsistency when we derive $\neg Con_R$ from it. R doesn't really *prove* (in the ordinary sense of establish-as-true) its own inconsistency, since we don't accept the theory as correct on the standard interpretation! That's why we used scare quotes in stating Theorem 74.

19.2   There are provable consistency sentences

We now turn to discuss ways in which, despite the Second Theorem, consistent theories can 'prove their own consistency'.

Assume all along that we are dealing with a theory satisfying the usual conditions, with a Gödel-numbering scheme in place. Roughly what we do is tweak the predicate Prf_T which canonically captures Prf_T to get another predicate CPrf_T which still expresses and captures Prf_T (assuming T is consistent), form a new provability predicate CProv_T from *that*, and use this to define a new consistency sentence CCon. We then show that T can (quite trivially) prove CCon_T – in fact, even Q can prove that, for any T.

(a)   First step. Suppose we put

Defn. 61. $\mathsf{CPrf}_T(\mathsf{x}, \mathsf{y}) =_{\mathrm{def}} \mathsf{Prf}_T(\mathsf{x}, \mathsf{y}) \wedge (\forall \mathsf{w} \leq \mathsf{x}) \neg \mathsf{Prf}_T(\mathsf{w}, \overline{\ulcorner \bot \urcorner})$.

Assuming T is consistent, we can then easily show that CPrf_T expresses and captures Prf_T. We just use a very similar argument to the one in §15.1 that shows that the very similar Rosserized predicate RPrf_T also expresses and captures Prf_T.

Next, predictably, we define

Defn. 62. $\mathsf{CProv}_T(\mathsf{x}) =_{\mathrm{def}} \exists \mathsf{v} \mathsf{CPrf}_T(\mathsf{v}, \mathsf{x})$

So the code number for a wff satisfies the revised provability predicate $\mathsf{CProv}_T(\mathsf{x})$ just if there is a T-proof of that wff, while there is no smaller-numbered proof of absurdity.

(b)   Now, there does seem to be some real motivation for being interested in such a provability predicate. For consider the following line of thought.

When trying to establish an as-yet-unproved conjecture, mathematicians will use any tools to hand, bringing to bear whatever background assumptions that they are prepared to accept in the context. The more improvisatory the approach, the less well-attested the assumptions, then the greater the risk of lurking inconsistencies emerging, requiring our working assumptions to be revised. We should therefore ideally keep a running check on whether apparent new results cohere with secure background knowledge. Only a derivation which passes the coherence test has a chance of being accepted as a kosher *proof.* So how might we best reflect something of the idea that a genuine proof should be (as we might put it) *consistency-minded*, i.e. should come with a certificate of consistency with what's gone before?

We might reasonably suggest: there is a consistency-minded proof of φ in the axiomatized formal system T iff (i) there is an ordinary T-derivation of φ with super g.n. m, while (ii) there isn't already a T-derivation of absurdity with a code number less than m. That's the idea which is formally reflected in the provability predicate CProv.

(c)   Now, lets define a corresponding consistency sentence

Defn. 63. $\mathsf{CCon}_T =_{\mathrm{def}} \neg\mathsf{CProv}_T(\ulcorner\bot\urcorner)$,

Since CProv_T expresses a provability relation, CCon_T (at first sight) 'says' that the relevant T is consistent. But it is immediate that

Theorem 76. $\mathsf{Q} \vdash \mathsf{CCon}_T$.

Proof. Arguing inside Q, the logic gives us $\neg(\mathsf{Prf}_T(\mathsf{a},\ulcorner\bot\urcorner)\wedge\neg\mathsf{Prf}_T(\mathsf{a},\ulcorner\bot\urcorner)$, where a is a dummy name. That easily entails $\neg(\mathsf{Prf}_T(\mathsf{a},\ulcorner\bot\urcorner)\wedge(\forall\mathsf{v}\leq\mathsf{a})\neg\mathsf{Prf}_T(\mathsf{v},\ulcorner\bot\urcorner))$, since Q can handle the bounded quantifier. Apply universal quantifier introduction to conclude CCon_T.   ⊠

(d)   What can we learn about the consistency of a theory T from last theorem? Predictably nothing at all!

Take T to be any theory of interest, e.g. Zermelo-Fraenkel set theory. Now, if T is indeed consistent, CProv_T indeed expresses provability-in-T, and CCon_T 'says' that T is consistent. But if T isn't consistent, then CProv_T doesn't always express that proof relation (suppose φ is provable but every proof of it has a larger g.n. than a proof of absurdity; then we won't have $\mathsf{CProv}_T(\ulcorner\varphi\urcorner)$). So if T isn't consistent, CCon_T can't necessarily be read as expressing consistency. *But this means that we need to know whether T is consistent before we can interpret what we've proved in deriving CCon_T.* Therefore the proof can't tell us whether T is consistent![1]

19.3   The 'intensionality' of the Second Theorem

The provability of Theorem 76 tells us nothing about the consistency of the relevant theory T. But it *does* tell us something important about what it takes to prove the Second Theorem.

Recall, using the box shorthand, we showed that if the HBL derivability conditions hold for the provability predicate Prov_T, we can't derive in T the consistency sentence Con_T (i.e. we can't derive $\neg\mathsf{Prov}_T(\ulcorner\bot\urcorner)$). Since we can derive

---

[1]It might help to think about this related example. Take a theory T. Let S be T's largest consistent sub-theory (we can make that idea respectable in various ways!). Suppose T is in fact consistent. Then S is none other than T! Does this mean, in this case, that the trivial observation that S is consistent is a proof that T is consistent?

in T the sentence CCon_T (i.e. we can derive $\neg\mathsf{CProv}_T(\ulcorner\bot\urcorner)$), the derivability conditions can't all apply to CProv_T. In fact, in showing that the derivability conditions do hold for Prov_T, we have to rely on the fact that Prov_T is defined in terms of a predicate expression Prf_T which *canonically* captures Prf_T. Putting it roughly, we need $\mathsf{Prf}_T(\overline{m},\overline{n})$ to reflect the details of what it takes for m to code a proof of the wff with g.n. n, and no more. Putting it even more roughly, we need Prf_T to have the right intended *meaning*. For if we doctor Prf_T to get another wff CPrf which still captures Prf_T but in a 'noisy' way, then – as we saw – the corresponding CProv need not satisfy the derivability conditions.

So we have the following important difference between the First and the Second Theorem. Suppose T is p.r. axiomatized and contains Q:

1. Take *any* old wff Prf_T^* which captures Prf, in however 'noisy' a way. Form the corresponding proof predicate Prov_T^*. Take a fixed point γ for $\neg\mathsf{Prov}_T^*(\mathsf{x})$. Then we can show, just as before, that $T \nvdash \gamma$, assuming T is consistent, and $T \nvdash \neg\gamma$ assuming ω-consistency as well. (The proof depends just on the fact that Prf_T^* captures Prf_T, not at all on *how* it does the job.)

2. By contrast, if we want to prove $T \nvdash \mathsf{Con}_T$ (assuming T is consistent and has a smidgin of induction), we have to be a lot more picky. We need to start with a wff Prf_T which does 'have the right meaning', i.e. which *canonically* captures Prf_T, and then form the canonical consistency sentence Con_T from *that*. It we capture Prf_T in the wrong way, using a doctored Prf_T^*, then the corresponding Con_T^* may indeed be provable after all.

Solomon Feferman describes results which depend on the open wff Prf_T "more fully express[ing] the notion involved" as 'intensional'.[2] The label carries baggage which we needn't worry about. The important thing is the easy-to-miss distinction Feferman wants to emphasize, between what it takes to prove the Second Theorem as opposed to the First.

---

[2] In his 'Arithmetization of Metamathematics in a General Setting', *Fundamenta Mathematicae* (1960), p. 35.

Further reading

And here, these notes will end (as did the lectures they were originally written to accompany). But let me mention three books which take the story further, and also give a pointer to a resource which suggests additional references.

First, though, a general remark. We have seen that Gödel's 1931 proof of his incompleteness theorem uses facts about primitive recursive functions: and these functions are a subclass, but only a subclass, of the effectively computable numerical functions. A more general treatment of the effectively computable functions (capturing *all* of them, if we are to believe Church's Thesis) was developed in the mid–1930s, and this in turn throws more light on the incompleteness phenomenon. So there's a choice to be made. Do you look at things in roughly the historical order, first encountering just the primitive recursive functions and learning how to prove initial versions of Gödel's incompleteness theorem before moving on to look at the general treatment of computable functions? Or do you do some of the general theory of computation first, turning to the incompleteness theorems later?

In these notes we have very largely taken the first route. You can explore this further in *IGT2*, i.e.

> Peter Smith, *An Introduction to Gödel's Theorems* (CUP, 2nd edition 2013).

A corrected version is now freely downloadable from https://logicmatters.net/igt, where there are also supplementary materials of various kinds.

Alternatively, you can take the second route by following

> Richard Epstein and Walter Carnielli, *Computability: Computable Functions, Logic, and the Foundations of Mathematics* (Wadsworth 2nd edn. 2000: Advanced Reasoning Forum 3rd edn. 2008)

This explores general computability theory first. It is a very good introductory text on the standard basics, particularly clearly and attractively done, with a lot of interesting and illuminating historical information too in Epstein's 28 page timeline on 'Computability and Undecidability' at the end of the book.

Those first two books should be very accessible to those without much mathematical background: but even more experienced mathematicians should appreciate the careful introductory orientation which they provide. Then, going up just half a step in mathematical sophistication, we arrive at a really beautiful book:

George Boolos and Richard Jeffrey, *Computability and Logic* (CUP 3rd edn. 1990).

This is a modern classic, wonderfully lucid and engaging. There are in fact later editions – heavily revised and considerably expanded – with John Burgess as a third author. But I know that I am not the only reader to think that the later versions (excellent though they are) do lose something of the original book's famed elegance and individuality. Still, whichever edition comes to hand, do read it! – you will learn a great deal in an enjoyable way.

For many more references, see the relevant sections of the Logic Study Guide, which can be downloaded from https://logicmatters.net/tyl.

Index of definitions

$\vDash$, 5
$\vdash$, 5
$\bar{n}$, 15
$\boxtimes$, 18
$\leq$, 38
Δ_0, 47
Π_1, 48
Σ_1, 48
$\vec{x}$, 57
β-function, 68
$\exists!$, 73
$\ulcorner \; \urcorner$, 79
ω-incomplete, 87
ω-inconsistent, 87
$\bot$, 116
$\Box_T$, 124

Ackermann function, 96

BA, 30
Baby Arithmetic, 30
bounded quantifiers, 47

canonically capture, 74, 132
canonically express, 71
capture
    function, 72, 98
    property, 23
    relation, 23, 72
CCon, 131
characteristic function, 65
Church's Thesis, 95, 106
complete logic vs complete theory, 6,
        14
completeness theorem, 6
    vs incompleteness theorem, 14
composition of functions, 55, 58
computable function, *see* effectively
        computable function

Con, 116
consistency sentence, 117
consistency-minded proof, 131
consistent theory, 11
correctly decides, 35
Curry's Paradox, 112

decidable theory, 21, 106
decides, *see* formally decides
definition by primitive recursion, 55, 57
definition chain for p.r. function, 59
definition of truth, *see* truth theory
derivability conditions, 124
diag, 80, 98
Diag, 99
diagonal function, 64
diagonalization, 26, 80
diagonalization lemma, 99
diagonalize out, 64
'do until' loop, 63

effectively
    axiomatized theory, 5
    computable function, 60, 63
    decidable, 2
    determinable, 2
    formalized language, 4
Entscheidungsproblem, 106
express
    function, 16
    property, 16, 23
    relation, 16

finitary mathematics, 121
first incompleteness theorem, 9
    formalized, 117
    Gödel's version, 92
    semantic version, 84, 85

135

semantic vs syntactic, 12–13, 94
   syntactic version, 88–90
first-order vs second-order quantifiers,
   41
fixed point, 100
'for' loop, 61
formalized language, 3
formalized first theorem, 117
formally decides, 6
formally undecidable, 6

G, G_T, 19, 83
Gödel number, 16, 76
   super, 76
Gödel sentence, 19, 83
   canonical, 83
   canonical vs wide sense, 101
Gödel-numbering scheme, 16, 75
   normal, 78
'Goldbach type', 92
Goodstein function, 96

Hilbert's Programme, 121

identity function, 58
'iff', 2
incompletability, 13, 86, 90
incompleteness theorem, see first, second incompleteness theorem
induction
   (informal) principle, 40
   (second-order) axiom, 41
   rule, 41
   schema, 41, 44
initial function, 56, 58
$I\Sigma_1$, 50

L_A, 34
L_B, 28
language of baby arithmetic, 28
language of basic arithmetic, 7, 34
   containing, 11
liar sentence, 20, 111
Löb's Theorem, 127
logicism, 8, 120

Master Argument for incompleteness,
   20, 114
minimization, 94

negation complete, 6
non-standard model
   of PA, 46
   of Q, 36

p.r., see primitive recursive
p.r. adequate, 92
p.r. axiomatized theory, 79, 95
PA, 43, 44
Peano Arithmetic, 43, 44
Prf, 17, 77
Prf, 18, 99
$Prfd$, 82
Prfd, 82, 99
primitive recursive
   function, 56, 58–59
   property, 65
   relation, 65
projection function, 58
Prov, 18, 100
provability predicate, 18, 100
   Rosser style, 108

Q, 35

recursive function, 95
recursively axiomatized theory, 95
recursively decidable, 105
Robinson Arithmetic, 35
Rosser sentence, 108
Rosser's Theorem, 108
RPrf, 107
RProv, 108

schema, 29
second incompleteness theorem, 118,
   124
   'intensionality', 132
semantics, 3
$Sent$, 17, 77
sound theory, 11
standard numerals, 7, 15
sufficiently strong theory, 24
syntax, 3
   arithmetization of, 17, 75

Tarski's theorem, 111, 114
term, 28
truth
   inexpressibility of, 114

undefinability of, 111
truth predicate
   formal, 111
   informal, 110
truth theory, 111

undecidable sentence, 6

Wff, 17, 77

ZFC, 117

CPSIA information can be obtained
at www.ICGtesting.com
Printed in the USA
BVHW052040200222
629616BV00005B/231